DEEP LEARNING IN VISUAL COMPUTING AND SIGNAL PROCESSING

DEEP LEARNING IN VISUAL COMPUTING AND SIGNAL PROCESSING

Edited by
Krishna Kant Singh, PhD
Vibhav Kumar Sachan, PhD
Akansha Singh, PhD
Sanjeevikumar Padmanaban, PhD

AAP | APPLE ACADEMIC PRESS

First edition published 2023

Apple Academic Press Inc.
1265 Goldenrod Circle, NE,
Palm Bay, FL 32905 USA
760 Laurentian Drive, Unit 19,
Burlington, ON L7N 0A4, CANADA

CRC Press
6000 Broken Sound Parkway NW,
Suite 300, Boca Raton, FL 33487-2742 USA
4 Park Square, Milton Park,
Abingdon, Oxon, OX14 4RN UK

Library and Archives Canada Cataloguing in Publication

Title: Deep learning in visual computing and signal processing / edited by Krishna Kant Singh, PhD, Vibhav Kumar Sachan, PhD, Akansha Singh, PhD, Sanjeevikumar Padmanaban, PhD.
Names: Singh, Krishna Kant (Telecommunications professor), editor. | Sachan, Vibhav Kumar, editor. | Singh, Akansha, editor. | Sanjeevikumar, Padmanaban, 1978- editor.
Description: First edition. | Includes bibliographical references and index.
Identifiers: Canadiana (print) 20220213763 | Canadiana (ebook) 20220213887 | ISBN 9781774638705 (hardcover) | ISBN 9781774638712 (softcover) | ISBN 9781003277224 (ebooks)
Subjects: LCSH: Deep learning (Machine learning) | LCSH: Signal processing. | LCSH: Three-dimensional imaging. | LCSH: Computer vision.
Classification: LCC Q325.73 .D44 2023 | DDC 006.3/1—dc23

Library of Congress Cataloging-in-Publication Data

CIP data on file with US Library of Congress

ISBN: 978-1-77463-870-5 (hbk)
ISBN: 978-1-77463-871-2 (pbk)
ISBN: 978-1-00327-722-4 (ebk)

About the Editors

Krishna Kant Singh, PhD, is working as Professor and Head, Department of CSE, Faculty of Engineering and Technology, Jain (Deemed-to-be University), Bengaluru, India. He is also the NBA coordinator for the deparment. He has wide teaching and research experience. Dr. Singh has acquired BTech, MTech, and PhD (IIT Roorkee) in the area of image processing and machine learning. He has authored more than 100 research papers in Scopus and SCIE indexed journals of repute. He has also authored 25 technical books. He is also an associate editor of *IEEE Access* (SCIE Indexed) and a guest editor of *Microprocessors and Microsystems, Wireless Personal Communications, and Complex & Intelligent Systems.* He is also member of editorial board of *Applied Computing and Geoscience* (Elsevier). Dr. Singh is an active researcher in the field of Machine Learning, Cognitive Computing, and 6G and beyond networks.

Vibhav Kumar Sachan, PhD, is a Professor and Additional Head of the Department of Electronics and Communication Engineering Department at the KIET Group of Institutions, Ghaziabad, Uttar Pradesh, India. Dr. Sachan started his career as a Lecturer of Electronics and Communication Engineering at the KIET Group of Institutions, Ghaziabad, Uttar Pradesh, India. During his academic career of 18 years, Dr. Sachan has taught various subjects at the undergraduate and postgraduate levels and has authored books, edited several conference proceedings, and written book chapters. He has published many papers in reputed international and international journals and conferences. He is on the editorial boards of several international and national journals. Dr. Sachan has received the Dronacharya Award & Letters of Appreciation for his contributions as Additional Head of the Department and Coordinator of Administrative Assignments by the Director General of KIET Group of Institutions, Ghaziabad, and Director of KIET Group of Institutions, Ghaziabad, respectively. Dr. Sachan holds BTech (with distinction), MTech (with distinction), and PhD degrees. He has been a recipient of merit scholarships throughout his education.

Akansha Singh, PhD, is working as Associate Professor in the School of Computer Science Engineering and Technology, Bennett University, India. She has to her credit more than 70 research papers, 20 books, and numerous conference papers. She has been the editor for books on emerging topics with publishers like Elsevier, Taylor and Francis, Wiley, etc. Dr. Singh has served as a reviewer and technical committee member for multiple conferences and journals of high repute. She is also an associate editor for the journals *IEEE Access* and *IET Image Processing*. She has also served as guest editor for several journals, including *Complex and Intelligent Systems*, Real Time Imaging, etc. Dr. Singh has also undertaken a government-funded project as principal investigator. Her research areas include image processing, remote sensing, IoT, and machine learning. Dr. Singh has a BTech, MTech, and PhD in Computer Science. She received her PhD from IIT Roorkee in the area of image processing and machine learning.

Sanjeevikumar Padmanaban, PhD, (Senior Member, IEEE) received the bachelor's degree in Electrical Engineering from the University of Madras, Chennai, India, in 2002, the master's degree (Hons.) in Electrical Engineering from Pondicherry University, Puducherry, India, in 2006, and the PhD degree in Electrical Engineering from the University of Bologna, Bologna, Italy, in 2012. He was an associate professor with VIT University from 2012 to 2013. In 2013, he joined the National Institute of Technology, India, as a faculty member. In 2014, he was invited as a visiting researcher with the Department of Electrical Engineering, Qatar University, Doha, Qatar, funded by the Qatar National Research Foundation (Government of Qatar). He continued his research activities with the Dublin Institute of Technology, Dublin, Ireland, in 2014. Further, he served as an associate professor with the Department of Electrical and Electronics Engineering, University of Johannesburg, Johannesburg, South Africa, from 2016 to 2018. Since 2018, he has been a Faculty Member with the Department of Energy Technology, Aalborg University, Esbjerg, Denmark.

Contents

Contributors

R. Ganesh Babu
Department of Electronics and Communication Engineering, SRM TRP Engineering College, Trichy, Tamil Nadu, India

Shalini Bhaskar Bajaj
Department of Computer Science, Amity University, Haryana, India

Arun Balodi
Atria Institute of Technology, Bangalore, India

Sunny Arokia Swamy Bellary
Charlotte, NC 28262, USA

Piyush Chauhan
School of Computer Science, University of Petroleum & Energy Studies, Dehradun, India

Pushpa Choudhary
Department of Information Technology, G. L. Bajaj Institute of Technology and Management, Greater Noida, India

Shilpa Choudhary
Department of Electronics and Communications, G.L. Bajaj Institute of Technology and Management, Gautam Budh Nagar, Greater Noida, India; E-mail: shilpadchoudhary@gmail.com

Kusumika Krori Dutta
M.S. Ramaiah Institute of Technology, MSR Nagar, Bengaluru, Karnataka 560054, India

Arush Jain
Department of Information Technology, G. L. Bajaj Institute of Technology and Management, Greater Noida, India

Aman Jatain
Department of Computer Science, Amity University, Haryana, India;
E-mail: amanjatainsingh@gmail.com

M. Jayashree
Anna University Regional Campus, Coimbatore, Tamil Nadu, India; E-mail: shrihema17@gmail.com

Utku Kose
Suleyman Demirel University, Isparta/Turkey

G. Manikandan
Department of Electronics and Communication Engineering, Dr. M.G.R. Educational and Research Institute, Chennai, Tamil Nadu, India

V. Megala
SRM Institute of Science and Technology, Ramapuram Campus, Chennai, Tamil Nadu, India; megala95@gmail.com

A. Nedumaran
Department of Electrical and Computer Engineering, Kombolcha Institute of Technology-Wollo University, Ethiopia

R. K. Pongiannan
SRM Institute of Science and Technology, Katankulathur, Chennai, Tamil Nadu, India;
E-mail: rkpannan@gmail.com

Akash Rajak
KIET Group of Institutions, 201206 Ghaziabad, India; E-mail: akashrajak@gmail.com

R. Selvameena
Department of Computer Science and Engineering, Dr. M.G.R. Educational and Research Institute, Chennai, Tamil Nadu, India

Bharti Sharma
DIT University, Dehradun, India

Durgansh Sharma
School of Computer Science, University of Petroleum & Energy Studies, Dehradun, India

Ajay Kumar Shrivastava
KIET Group of Institutions, 201206 Ghaziabad, India; E-mail: ajay@kiet.edu

Vidit Shukla
Department of Electronics and Communications, G.L. Bajaj Institute of Technology and Management, Greater Noida, India

Roohi Sille
School of Computer Science, University of Petroleum & Energy Studies, Dehradun, India;
E-mail: rsille@ddn.upes.ac.in

Arun Kumar Singh
Department of Information Technology, G. L. Bajaj Institute of Technology and Management, Greater Noida, India

Akansha Singh
School of CSET, Bennett University, Greater Noida, India

Krishna Kant Singh
Faculty of Engineering & Technology, Jain (Deemed-to-be University), Bengaluru

Poornima Sridharan
Anna University, CEG Campus, Chennai, Tamil Nadu, India; E-mail: poojeyaani@gmail.com

Kashish Srivastava
G. L. Bajaj Institute of Technology and Management, Gautam Budh Nagar, Uttar Pradesh, India;
E-mail: kuhusri2121@gmail.com

Arun Kumar Tripathi
KIET Group of Institutions, 201206 Ghaziabad, India; E-mail: mailtoaruntripathi@gmail.com

Ashish Tripathi
Department of Information Technology, G. L. Bajaj Institute of Technology and Management, Greater Noida, India

Khushboo Tripathi
Department of Computer Science, Amity University, Haryana, India

Shraddha Upadhaya
Department of Information Technology, G. L. Bajaj Institute of Technology and Management, Greater Noida, India

Prem Chand Vashist
Department of Information Technology, G. L. Bajaj Institute of Technology and Management, Greater Noida, India

Vidushi
KIET Group of Institutions, 201206 Ghaziabad, India; E-mail: vidushi.mtech@gmail.com

Abbreviations

AI	artificial intelligence
ANN	artificial neural network
API	application programming interface
ASDs	antiseizure drug
ASR	automatic speech recognition
AWF	adaptive Weiner filter
BP	back propagation
BPTT	backpropagation through time
CAD	computer-aided-design
CBIR	content-based image retrieval
CM	confusion matrix
CNNs	convolutional neural networks
CPU	central processing unit
CS	chitosan
CSF	cerebrospinal fluid
CT	computed tomography
CVPR	Computer Vision and Pattern Recognition
DBN	deep belief network
DCNN	deep convolutional neural network
DDL	distributed deep learning
DL	deep learning
DNNs	deep neural networks
DRL	deep reinforcement learning
DSC	dice score coefficient
ECG	electrocardiogram
EEG	electroencephalography
ELM	extreme learning machine
EMFSE	expert maximum fuzzy-sure entropy
EMG	electromyogram
EOG	electrooculogram
FC	fully connected
FLOPs	floating point operations
GAN	generative adversarial network

GM	gray matter
GPU	graphics processing units
GRU	gated recurrent units
GUI	graphic user interface
HGG	high grade gliomas
HNF	human nephron filter
HOG	histogram of oriented gradients
ILSVRC	ImageNet Large Scale Visual Recognition Challenge
IoT	internet of things
LDA	linear discriminant analysis
LRF	local receptive field
LSTM	long short-term memory
MCC	Matthew's Correlation Coefficient
mIoU	mean of intersection of union
ML	machine learning
MLPNN	multilayer perceptron neural network
MNIST	Modified National Institute of Standards and Technology
MRI	magnetic resonance imaging
NIAC	NASA's Innovative Advanced Concepts
NREM	nonrapid eye movement sleep
NS	neutrosophic set
PCA	principal component analysis
PVC	plasticized polyvinyl chloride
RBM	restricted Boltzmann machine
RBS	ribosomal restricting locales
RMSE	root mean square
RNN	recurrent neural network
ROI	region of interest
RPN	regional proposal network
RST	renal substitution treatment
SAEs	stacked auto-encoder
SGD	stochastic gradient descent
SGDM	Stochastic gradient descent momentum
SIFT	scale invariant feature transformation
SSD	single shot detector
SSEP	steady-state evoked potentials
TCIA	The Cancer Imaging Archives
tFNA	tetrahedral system nucleic corrosive

TMS	transcranial attractive incitement
VJ	Viola-Jones
WM	white matter
YOLO	you only look once

Preface

The edited book, *Deep Learning in Visual Computing and Signal Processing,* discusses the applications and challenges of deep learning for visual computing and signal processing. This book covers the fundamentals and advanced topics in designing and deploying techniques using deep architecture. It can serve as a guide for researchers, engineers, and students who want to have a quick start on learning and/or building deep learning systems. This book provides a good theoretical and practical understanding and complete information and knowledge required to understand and build deep learning models from scratch. The book focuses explicitly on deep learning, filtering out other material that co-occur in many deep learning books on visual computing and signal processing. This book also presents fundamental and advanced concepts of deep learning and its application in visual computing and signal processing.

This book is a resource that can help readers enter and master the field of deep learning. The book focuses on the usage of deep learning in the field of visual computing and signal processing. The book is an amalgamation of deep learning concepts with visual computing and signal processing applications. It will provide readers a comprehensive knowledge of the field. This book provides information starting with the fundamental to the latest research being done in the field.

There are several books available in deep learning and machine learning. There is no such book available that focuses on the usage of deep learning specifically for visual computing and signal processing. It also provides the research applications of deep learning in visual computing and signal processing. Thus, this book is unique in terms of the topics and related contents it covers. Readers from many domains will be interested as it covers three major fields. Also, it will be appealing for the readers who tend to research in this field as the book covers latest research topics.

CHAPTER 1

Deep Learning Architecture and Framework

ASHISH TRIPATHI[1*], SHRADDHA UPADHAYA[1], ARUN KUMAR SINGH[1], KRISHNA KANT SINGH[2], ARUSH JAIN[1], PUSHPA CHOUDHARY[1], and PREM CHAND VASHIST[1]

[1]*Department of Information Technology, G. L. Bajaj Institute of Technology and Management, Greater Noida, India*

[2]*Faculty of Engineering & Technology, Jain (Deemed-to-be University), Bengaluru, India*

[*]*Corresponding author. E-mail: ashish.mnnit44@gmail.com*

ABSTRACT

Nowadays, deep learning (DL) is one of the most trending technologies in the world. It is a subset of machine learning in artificial intelligence. It is very helpful while working with unlabeled and unstructured data for unsupervised learning. DL is concerned with several algorithms that are inspired by the artificial neural network. It provides the best results in the area of image processing as well as the natural language processing. There are several open-source frameworks that allow the implementation of the common DL algorithms; it enhances the growth of DL. The purpose of this work is to give a proper explanation of the DL architecture and framework. There are different varying and widely used architecture and algorithms are used in the DL. Long short-term memory and convolutional neural networks are the two oldest approaches that are mostly used

Deep Learning in Visual Computing and Signal Processing. Krishna Kant Singh, Vibhav Kumar Sachan, Akansha Singh & Sanjeevikumar Padmanaban (Eds.)

in various applications. The objective of this work is to know about the DL architecture, and frameworks that fulfill the requirements of the end-users. It includes all the DL-related information, principle, and motivation regarding learning algorithms for the deep structures.

1.1 INTRODUCTION

Today, deep learning (DL) is one of the most popular technologies in the world. In recent times, DL has been emerged as the most demanding technology due to its wide applicability and success rate in various application domains. The application of DL is growing very fast in several domains and has been resolving many real-world problems in the public interest. DL is also used in the new application areas where there is an immense need to find the solution for various issues. The DL method has been mainly categorized in deep supervised and unsupervised learning as we can see in the case of machine learning. But the DL methods have shown their outstanding performance in comparison to the traditional machine learning approaches. That is why DL is becoming more popular in the field of computer vision, machine translation, image processing, speech recognition, bioinformatics, medical imaging, and many others.

As we know, DL is the subset of machine learning (ML) that comes under artificial intelligence (AI) as shown in Figure 1.1. The main purpose of the DL is that it gives the closest result of the given problem. In ML, the main problem is choosing the right attributes when input cases are needed to solve a specific problem. As an alternative, learning features automatically with the help of deep structures. In DL structures, there is a different construction for many structures other than function and can be considered as an appropriate choice. Certainly, some nonlinear functions have more representation in the case of features with deeper architecture than others like support vector machine. For example, it has emerged that n-bit parity function can be encoded by a neural transmission network with the complexity of O (log n) hidden layers and complexity of O (n) neurons, while the feed forward network has only one hidden layer. It requires an emerging number of parallel neurons to perform the same task .[1] In addition, for highly validated tasks, learning algorithms are based on closed-loop algorithms that are more sensitive to the degradation of dimensions.[2] Now, right in depth the structure of the issue used a shared description (shared presentation) and similarly it can be a tangible thing to answer. Deep building training is intense. Great architectural skills are trained prior to the recognition of a knowledge-based basis for solving specific problems. For example, if there are a large number of layers in the neural network, backpropagation has less influence than the original layers. This causes the origin of the gene to penetrate between the plain and the local minima. It is the reason that the experts often look at the neural network limit on one or two hidden layers, to solve this problem before training in deep structures.

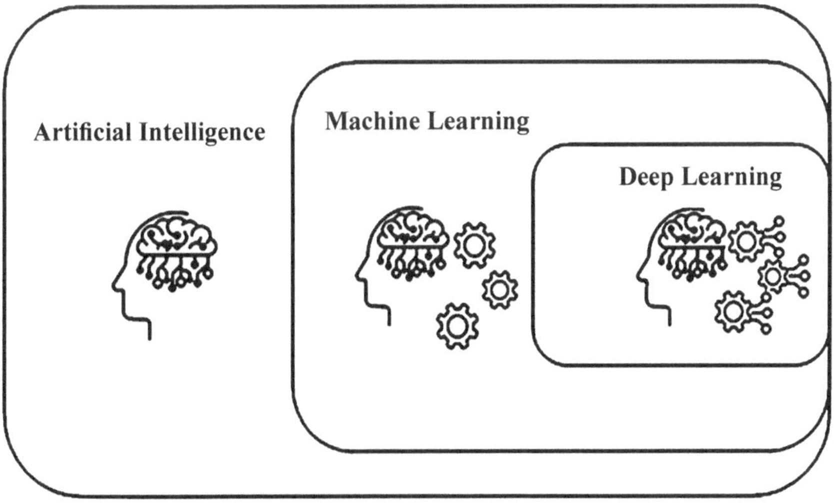

FIGURE 1.1 Connection of AI, ML, and DL.

In today's world, DL models are widely used by various industries. There are various industrial sectors and organizations that use different types of DL such as consumer products and medical technology. DL is highly preferred for complex problems such as natural language processing, image processing, and speech recognition as it provides additional accuracy of results.

DL works better for big data sets. DL models classify problems by looking at texts, sounds, images, and other parameters. It is the most common way to learn typical network architecture. DL does not require replanning and reproducing the best results. The result will be produced without reimagining the result. DL has the ability to identify patterns in detail and provide accurate and precise results. Some applications of smart devices use DL for automatic hearing and speech translation. Medical analysts use DL to detect cancer cells. For industrial purposes, it is used for risk management in operating heavy machinery. DL enables the use of algorithms to process data and copy the thinking process. It comes from the quality of dealing with ideas. DL uses different layers of the algorithm for data processing, speech recognition, and visualization.

1.1.1 DEEP LEARNING: A HISTORICAL BACKGROUND[3,4]

The history of DL began in 1943. In 1943, Walter Pitts and Warren McCulloch created a computer model that was based on the neural networks of the human brain. In this case, the combination of algorithms and mathematics used is called "thresher logic" to copy the thinking process. Since then, DL has improved slowly. A previous attempt to invent DL algorithms came from Alexey Grigoryevich Ivakhnenko and Valentin Grigor'evich Lapa in 1965. They used the models with polynomial opening functions and analyzed similar statistics. For each layer, the best features will be selected and moved to the next layer. It was a soft and learned process. In the 1970s, the first AI came into operation. It leads to limited funding deficits in this DL and AI research. Fortunately, some were doing this research at no cost. Kunihiko Fukushima used the first neural networks to verify.

It was built with multiple binding and authentication plates. In 1979, Fukushima developed an artificial neural network, called Neocognitron. It enables the computer to identify the visual patterns. A large number of Neocognitron ideas continue to be used. Modern Neocognitron not only reveals patterns with missing information but can also complement the image by adding missing details. It can be described as "humble." In the 1970s, backpropagation, the use of errors in training DL models, gradually increased. In 1989, YannLeCun combined the neural network available with backpropagation over learning "handwritten" numbers, and 1985–1990s was the year when the second winter of AI came in. It also affected the study of neural networks and DL. The next significant change in DL took place in 1999. By this time, computers were starting to accelerate and graphics processing units (GPU) were being developed. Currently, the emergence of AI and big data processing are both subject to DL. Figure 1.2 shows the performance of DL in comparison to the traditional learning algorithms.

1.1.2 DEEP LEARNING: OVERVIEW, DEFINITION, AND UNDERSTANDING

Machine learning is a technology that supports many different aspects of today's global society, social networks, and various e-commerce websites. It is also available in different consumer products such as cameras and smartphones. Various programs have been proposed using ML technology. These devices are used for object recognition, translating speech into text, comparing different posts or products, news items such as users' interests, and so on.

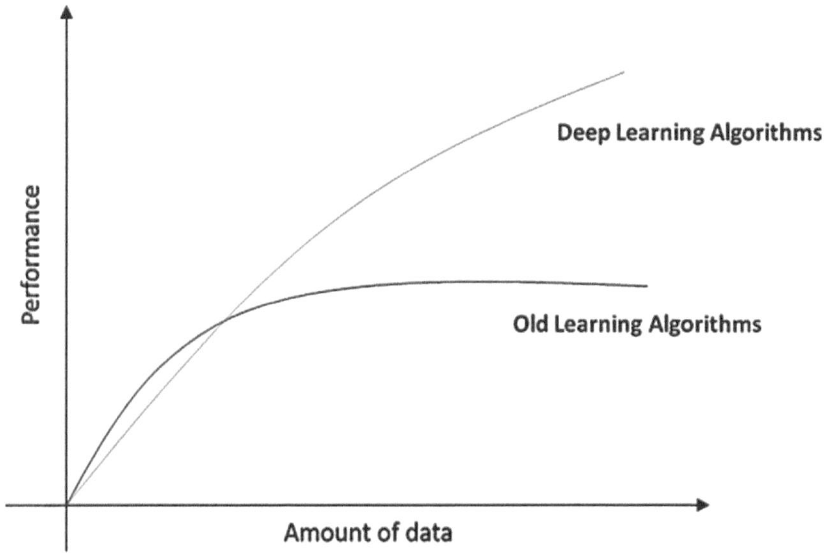

FIGURE 1.2 Performances of DL algorithms with respect to the older algorithms.

Use of these applications is increasing day by day. Increasingly, these systems are implementing a phase of the process called DL. Executive ML methods had their limitations for reducing raw natural data. The development of a machine learning system or pattern recognition program requires the attention engineering and background expertise. It is necessary to design a feature additive. This feature identifier is designed to convert raw data into an appropriate internal presentation or feature display. Using this representation learning, the domain determines or separates input patterns.

Presentation readings are a set of methods that allow the machine system to work with raw data. It naturally determines the representations required for group recognition or collection. Now, DL approaches are ways to learn representation. DL methods have many levels of representation, which are obtained by building simple but linear modules. Each level prepares the presentation to another level by symbolizing the other at the highest, excessive. With the help of those changes, we can learn more complex tasks. For the purposes of segmentation, higher layers of the presentations fetch information about the input.

This is the key to differentiation and the greater potential for nonlinearity.

For example, a process image comes in the form of similar pixel values, the initial presentation sometimes tells the presence and absence of edges somewhere. The second layer usually determines the patterns by recognizing the particular shape of the edges, despite the slight differences in the edges of the edges. The third layer may accumulate patterns in larger mixes that corresponds to the components of the adjacent elements, and future layers may find objects as cohesive of these components. The central idea of DL is that these layers of features are not designed by engineers. These are understood from the data by using a standard learning process approach.

For many years, DL played a vital role in solving problems that have endured the best efforts of the intelligence community. It provides excellent results in determining complex structures in high-quality data. Therefore, it is suitable for various fields such as science, government, and business.

In addition to incorporating track records in image recognition[5,6] and speech recognition,[7,8] it strikes another possible mechanism for machine learning in predicting drug molecule activity,[9] examining particle acceleration,[10] restructuring brain circuits,[11,12] and clarifies the effects of noncoding DNA mutations on disease type and disease.[13,14] Surprisingly, in-depth reading has positive effects of different tasks on natural language comprehension,[15] direct subject segmentation, auditory analysis, question answering,[16] and language translation.[17]

DL in advance will have many positive effects in the future because it requires a little bit of engineering on hand. So it can be easily exploited with an increasing amount of accessible data and easy calibration. New learning algorithms and structures, currently built on deep neural networks (DNNs) will only facilitate the development and progression of DL.

1.1.3 DEEP LEARNING: APPLICATIONS

There are various applications of DL that can be seen in the real world. Below some of them have been discussed.

1.1.3.1 AUTOMATIC MACHINE TRANSLATION [18, 19]

The DL helps to automatically translate the sentence of one language into another language. This machine translation technique is not new and it is in practice for a long time, but DL has gained wide acceptance and outstanding results, especially in automatic text translation and automatic image translation.

The translation is based on learning the dependencies of words and their mapping to the new language. In this whole process, it does not require preprocessing of the sequence of words.

1.1.3.2 IMAGE RECOGNITION [20,21]

Image recognition is another popular application of DL. Image recognition is widely used in various sectors such as retail, gaming, tourism, social media, and so on. The DL-based image recognition includes the identification and detection of any object in the image. It recognizes the object based on the context and nature of the content. The object classification within an image is based on the previously known set of objects, while object detection recognizes one or more objects within an image and draws a box around them.

The standard measurement set for image classification is MNIST (Modified National Institute of Standards and Technology) for data set. MNIST is made up of handwritten digits and includes 60,000 training examples and 10,000 exam examples. Having a small size, it allows users to test multiple configurations. A wide range of similar results is available.

1.1.3.3 MARKETING RESEARCH [22]

Regression and classification models of DL can be used in the analysis of marketing campaigns, market segmentation, market feasibility, and many more. However, DL will be only helpful in the presence of a huge amount of data to analyze the market trend. If the available data are less, we should go for the traditional machine learning algorithms.

1.1.3.4 AUTOMATIC TEXT GENERATION[23]

In automatic text generation, the new text is generated that is based on a learned text corpus. The new text generation occurs either character-wise or word-wise. The DL-based model contains all the required capability to learn and handle characters, punctuations, and even text style in the corpus. In this context, a recurrent neural network (RNN) is used to generate text. For this, RNN takes input strings sequences to learn the relationship between the items of the sequences. Recently, a great effort using long short-term memory (LSTM) RNN has been taken and used a character-based model to generate a character at a time. Recently, to ease the solution of automatic text generation, LSTM RNN-enabled character-based model has been demonstrated to generate a single character at a time.

1.1.3.5 OPTIMIZATION OF VISUAL ARTS [24]

As deeper learning improves the process of image recognition, that's why it expands the use of DL strategies in various works of art. DNNs have proven to be able to (a) identify the style time of a given drawing, (b) neural styles transfer—captures a given art style, and (c) produce a fun image based on random input types.

1.1.3.6 NATURAL LANGUAGES MODELS [25, 26]

Since the 2000s, neural networks are being used to generate language models. LSTM helped to improve machine translation and language imitation.

1.1.3.7 HEALTHCARE SECTOR [28]

In recent years, different applications like the diagnosis of critical diseases, that is, breast cancer, Alzheimer's disease, and so on, patient monitoring system, personalized medicine, drug manufacturing, and many more in the healthcare system have been supported by DL approaches.

DL is playing a significant role in optimizing the healthcare sector through supporting the innovation of developing medicines for existing and newly found diseases, health management, computer-based detection and diagnosis, medical imaging, smart decision support system, and so on.

Other important initiatives for DL include image restoration, text sentiment analysis, automatic handwriting recognition, customer relationship management, bioinformatics, medical image analysis, mobile advertising, financial fraud, automatic colorization, advertising, and many more.

1.1.4 CATEGORIES OF DEEP LEARNING TECHNIQUES

There are mainly two categories of the DL techniques, that is, deep supervised and unsupervised learning. Another category of DL is deep reinforcement learning (DRL). The details of these categories are as follows.

1.1.4.1 DEEP SUPERVISED LEARNING

It is the most commonly used type of machine learning. Supervised learning uses the labeled data set; it means the output is already known for the given input. In the training phase, the training sets are used as input data to train the model where each input has a label with its corresponding output. During the training, the model learns that what would be the output when the input comes. The scheme of deep-supervised learning is shown in Figure 1.3. In the case of deep supervised learning, we use some set of inputs and their corresponding outputs that can be represented as (In_i, O_i), where In_i represents the input and O_i is used for output. It forms a training set in the pairs of $\{(In_1, O_1), (In_2, O_2) \ldots (In_n, O_n)\}$. During the training, the system get loss value in predicting the output as $O_i' = f(In_i)$. This loss value is the difference between the predicted and actual result obtained from the trained model. Here, the received loss value can be represented as $l(O_i, O_i')$. To reduce the loss value, system modifies the network in an iterative manner to get the desired output with better accuracy. After the successful completion of the training process, the system becomes ready to give the desired response. Examples of deep supervised learning algorithms are convolutional neural network (CNN/ConvNet), DNN, LSTM, RNN, and gated recurrent unit (GRU).

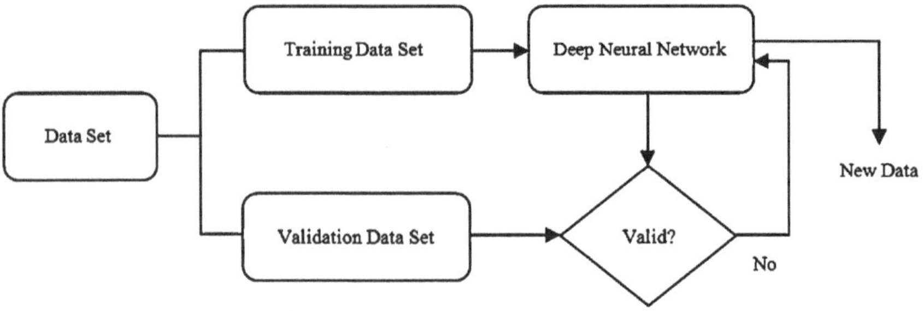

FIGURE 1.3 Scheme of deep supervised learning.

1.1.4.2 UNSUPERVISED DEEP LEARNING

Unsupervised learning does not use the labeled data set for training purposes and model creation. In the case of deep unsupervised learning, the system learns the significant features and unknown relationship among the input data and uses such type information to train the model. Unsupervised learning generally applies dimensionality reduction, generative approaches, and clustering to handle unlabeled data. The scheme of unsupervised DL is shown in Figure 1.4. The DL algorithms such as generative adversarial network, restricted Boltzmann machines (RBM), RNN, and auto-encoders support nonlinear dimensionality reduction and clustering. These algorithms are applicable for unsupervised learning in various application domains.

1.1.4.3 DEEP REINFORCEMENT LEARNING

DRL is the combined approach of reinforcement learning with deep neural network as shown in Figure 1.5. They jointly help the agent to take the best possible step to achieve the desired goal. Reinforcement learning is a subset of unsupervised learning. RL learns from the experience based on the reward received against the suitable action taken in a particular situation. The reinforcement agent decides what action should be taken to resolve the issue for a given task. Reinforcement learning does not use training data set as used in supervised learning. The experience gained by RL based on the outcome of the developed model. The model receives feedback in terms of reward or punishment that depends on its output. Such experience is used for further training purposes and to improve the performance of the model. The maximum reward decides the quality of the solution.

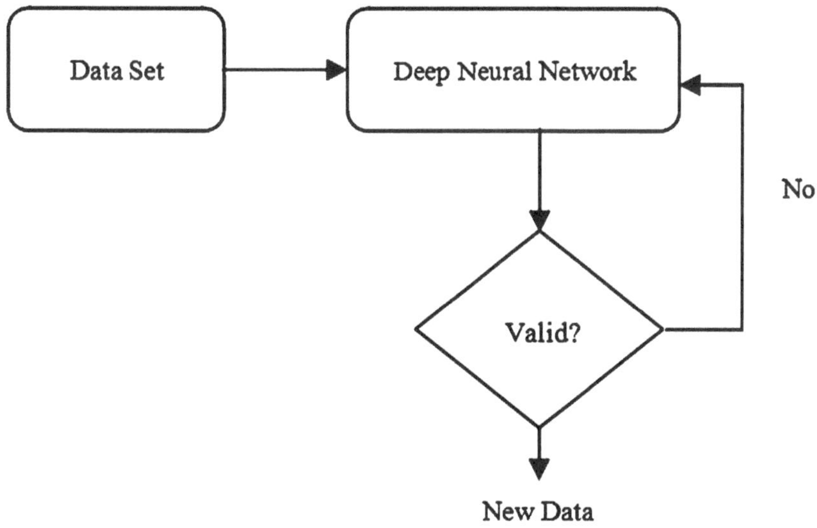

FIGURE 1.4 Scheme of deep unsupervised learning.

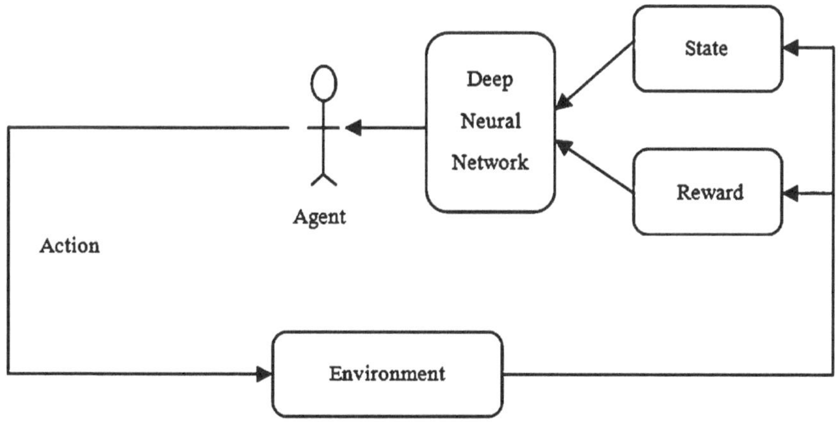

FIGURE 1.5 Scheme of deep reinforcement learning.

1.2 DEEP LEARNING: ARCHITECTURERCHITECTURE [29, 30]

In AI, unity structures are available for more than 70 years, but new structures and GPUs enable them to lead the way in AI. The past two decades have led to the development of DL that broadens the neural network's perception of the number and types of problems on a large scale. DL uses many different building blocks and many different algorithms. There are five DL structures for expansion in the last 20 years. Surprisingly, LSTM and CNN are two of the oldest methods.

Time-to-time different architectures of DL have been proposed; some of them are discussed below. Figure 1.6 shows the evolution of DL architecture.

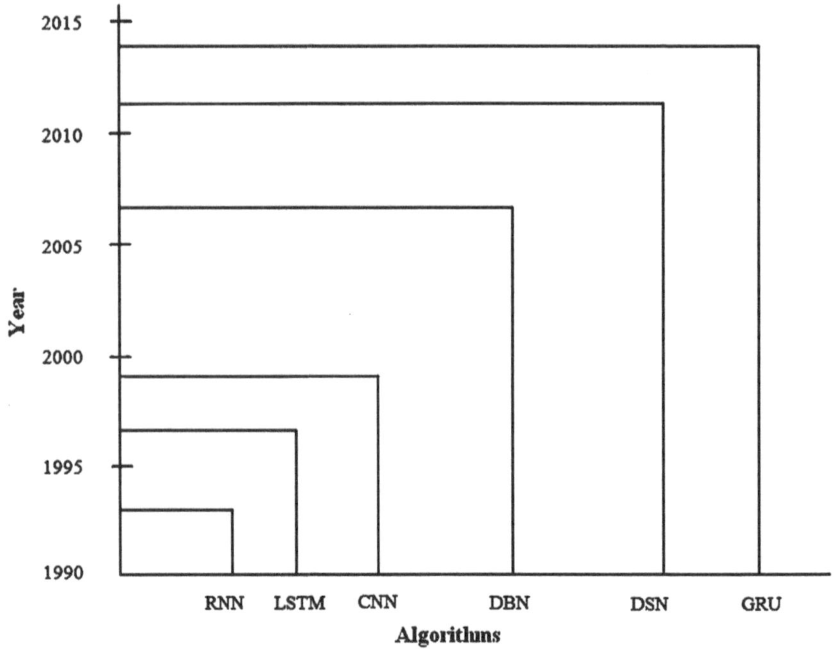

FIGURE 1.6 Evolution of the deep learning architecture.

1.2.1 RECURRENT NEURAL NETWORKS [31,32]

RNN was established on the basis of David Rumelhart's work in 1986. In 1982, John J. Hopfield used a particular type of RNN called Hopfield networks. In 1993, the neural history compressor program solved a "deepest learning" piece of work that required the following 1000 layers in a timely RNN. The term "neural network" is used loosely to refer to two broad categories of networks with a common relational structure, one with a short effect and the other with a lasting effect. Two phases of the network produce robust transient activation. RNN is a family of advanced neural networks. These are different from other networks due to the feed network, while the infinite feed network is a cyclic graphic technology. It is directed to send information from time to time. Leading researcher Jürgen Schmidhuber describes the RNN as follows.

The latest neural networks allow for computationally sequential and sequential computation, and in principle, it can include anything a traditional computer can combine. Unlike traditional computers, however, Recu neural networks are similar to the human brain, which is a large response network of interconnected neurons that can learn to translate a live input signal into a useful motor output sequence.

The mind is a wonderful example as it can solve many problems that modern machines cannot solve.

RNN is the basis of all DL structures. It is unique to the multilayer transmission network because of its connectivity between neurons. In this case, the neurons interact with the neuron of the other layers and the neuron of the same layer to provide feedback. Response connection supports RNN storing past input memory. The model creates problems over time.

In the past, it was difficult to train RNNs, but now, advances in research make them available to the doctor. This authorizes the network to continue to specify while moderating each-and-every windows input device. Measurement of time dimensions is a distinctive feature of RNNs. It encompasses a broad set of structures. The key identifier is the network response, which can be expressed in a hidden layer, an output layer, or a combination. The basic diagram of RNN is shown in Figure 1.7.

1.2.2 LONG SHORT-TERM MEMORY NETWORK [33,34]

LSTM was introduced by Schimdhuber and Hochreiter in 1997, but its popularity has increased in recent years. It is an RNN method used in the area of DL. LSTM is a special kind of RNN that has ability to learn long-term dependencies.

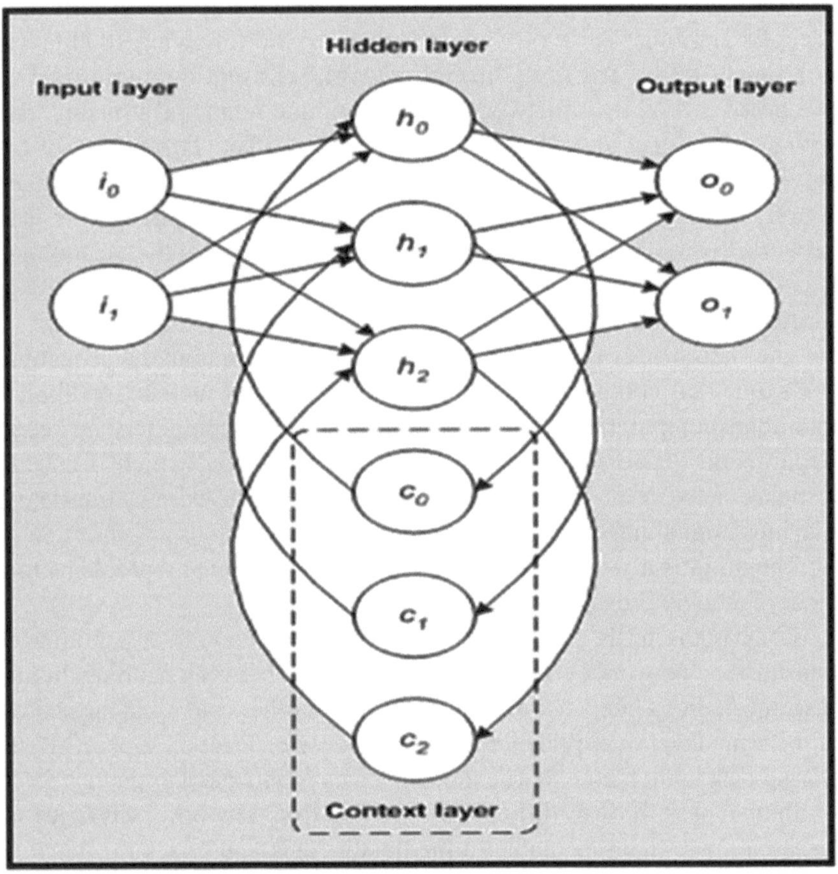

FIGURE 1.7 Recurrent neural network.

Unlike standard feeder neural networks, LSTM also has a feedback connection. It considers the sequence of all data by comparing it to a single data point. There are several functions by which LSTM work. LSTM is derived from the normal network structures formed by the neuron and instead introduced the concept of the memory cell. These memory cells can store their values temporarily or as long as the input function, allowing the cell to remember what's important and not just its current calculated value.

The LSTM memory cell has three gates. These gates command the flow of information to enter or exit the cell. A single input gateway controls the flow of new information into memory. When an existing piece is forgotten, the forgotten gate controls and allows the cell to remember the new data.

The output gate controls when the information contained in the cell is used to exit the cell. The cell also holds strings, which give birth to each gate. These weights are made larger based on the error of the network output using training techniques such as backpropagation through time (BPTT). BPTT is used to train certain types of RNN and it is a gradient-based technique. The working model of LSTM is illustrated in Figure 1.8.

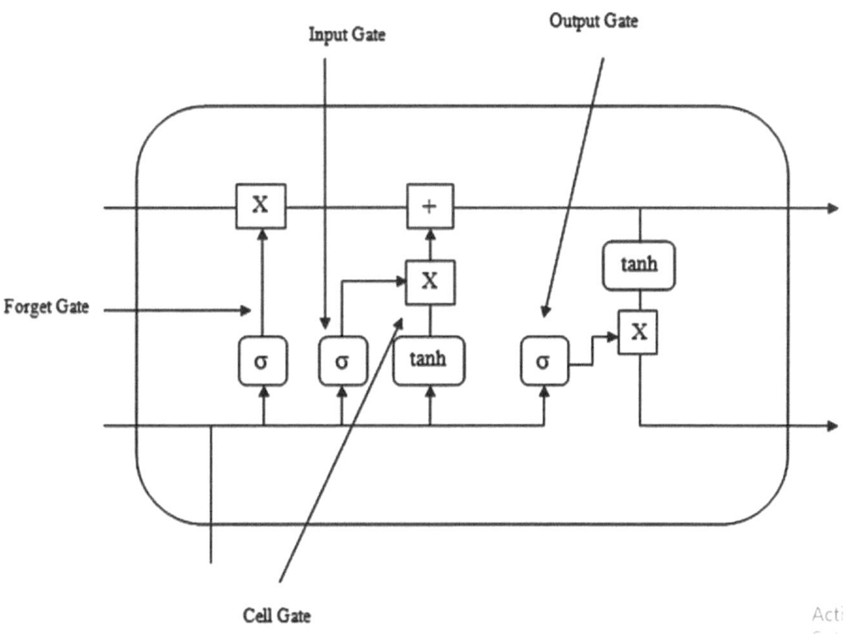

FIGURE 1.8 LSTM.

1.2.3 GATED RECURRENT UNITS [35, 36]

In 2014, a simple LSTM method was described and was called GRU. This model includes two gates. It helps to eliminate the output gateway that exists in the LSTM model.

GRU has the same functionality but has lesser parameters than LSTM. It has less weight results in faster performance. GRU has two gates: update gate and reset gate. These gates are used to determine which type of information required to be passed to the output. These gates can be trained to store the information for a long time. The update gate specifies how much previous information should be stored and passed along to the future. The reset gateway describes the association of new entries with previously stored information and also decides which past information needs to be forgotten. Compared to LSTM, GRU is simpler. It can be trained quickly and is efficient. However, LSTM can be more efficient and can result in better data quality.[1] Figure 1.9 shows the working model of GRU.

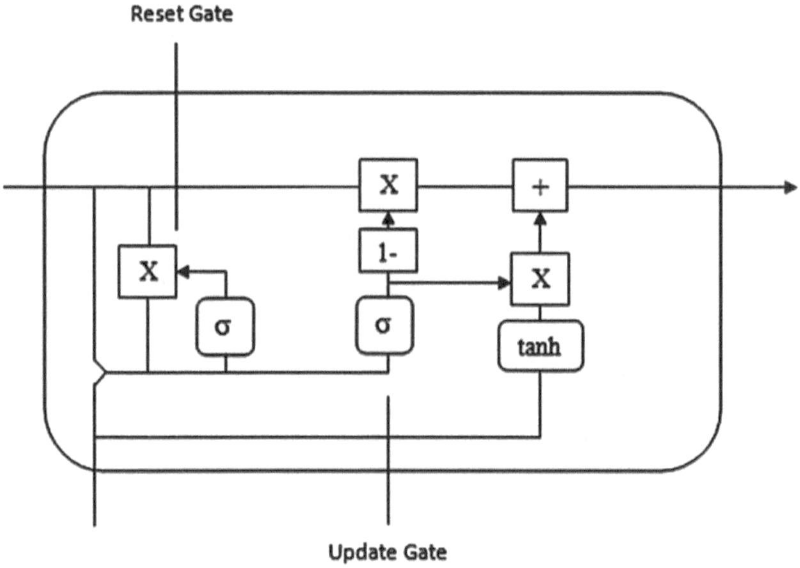

FIGURE 1.9 GRU.

1.2.4 CONVOLUTIONAL NEURAL NETWORKS (CNN) [37, 38]

CNN construction is very effective in image processing. The goal of CNN is to capture the qualities of high-order data through convolutions. This is well suited for item recognition with high-resolution image editing contests. CNN can recognize faces, road signs, and other aspects of sensitive data. It combines text analysis with visual character recognition. It is also important as it can analyze words as abstract text units.

CNN is also good at analyzing sounds. CNN is very similar to the internal neural network. It has three types of layers: an input layer, one or more hidden layers, and an output layer. It is different from the normal neural network. It creates an expectation that the input is imagery. It also allows the installation of sound "building blocks." It uses convolutional filters that convert 2D input data into 3D data. Figure 1.10 shows the architecture of CNN.

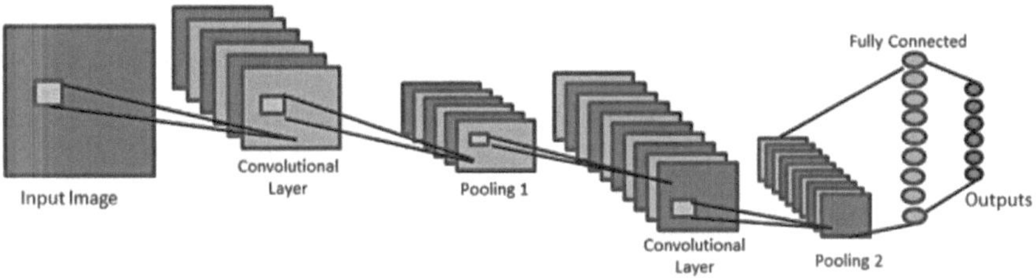

FIGURE 1.10 Architecture of CNN.

The LeNet CNN architecture feature consists of different layers that complete the extraction and then categorized as shown in Figure 1.11. The image is divided into corresponding areas composed of the optical layer. It captures elements from the input images. The next step in this process is collaboration. It reduces the size of the collected features. It preserves the most important information. Another step was taken to pull the straps and the joint together while feeding the fully integrated multilayer perceptual. And, the last layer is the output layer of this network. The network is trained using backpropagation.

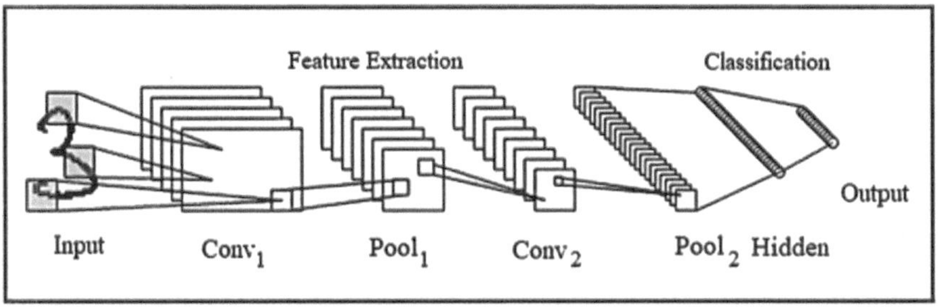

FIGURE 1.11 Architecture of LeNet.

1.2.5 DEEP BELIEF NETWORKS (DBN) [39,40]

Deep belief network (DBN) is a building block of good design but contains a novel training algorithm. All connected pairs are an RBM as shown in Figure 1.12. DBN is represented as a stack of RBMs. A deep neural network can be thought of as a stack of RBMs. This network involves training in two phases. These stages are preprepared and well-organized training. In the pretraining phase, each RBM is trained to configure its installation. This installation will be assigned to the next RBM. This process will continue until each layer is complete. The network will go into fine-tuning when the related phase is completed.

In the optimization phase, the output layer is used to mark the network. Major DBN programs are image recognition, site labeling, information retrieval, natural language processing, failure prediction, etc.

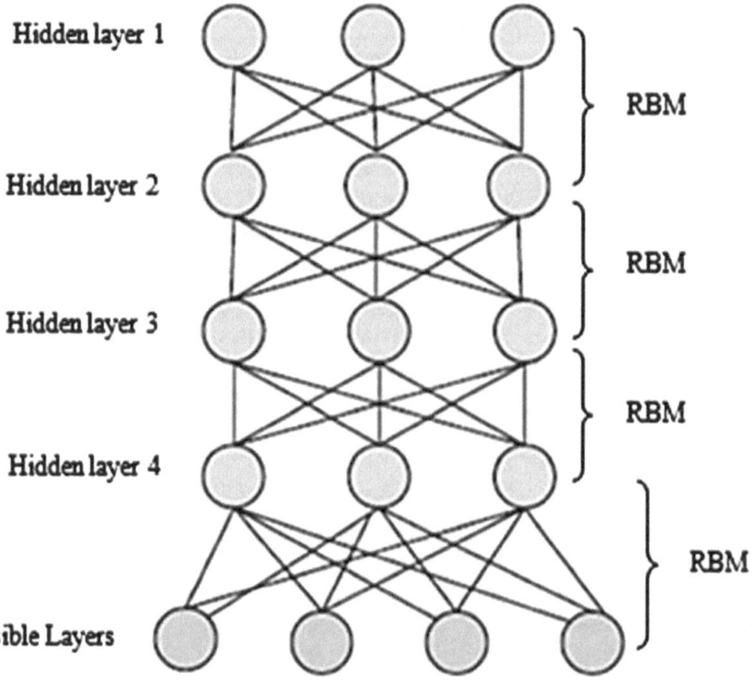

FIGURE 1.12 Deep belief network.

1.2.6 DEEP STACK NETWORKS (DSN) [41]

Deep stack network (DSN) is a storage structure as shown in Figure 1.13. It is also known as a deep convex network. This is contrary to the standard framework for DL. It is a deep set of distinct networks, each with its own hidden layers. It answers one of the deeper learning issues that is the complexity of training. Each layer of the DL structure increases the complexity of training. So, the DSN treats the training as a set of individual training problems rather than a single problem.

DSN consists of a set of modules. One of these modules is a low-level DSN function. There are three modules for DSN. Each module is made up of an input layer, hidden layer, and output layer. These modules are stacked on top of each other. The module contains input and output elements and a basic input vector. This type of structure allows the entire network to handle more complex partitions.

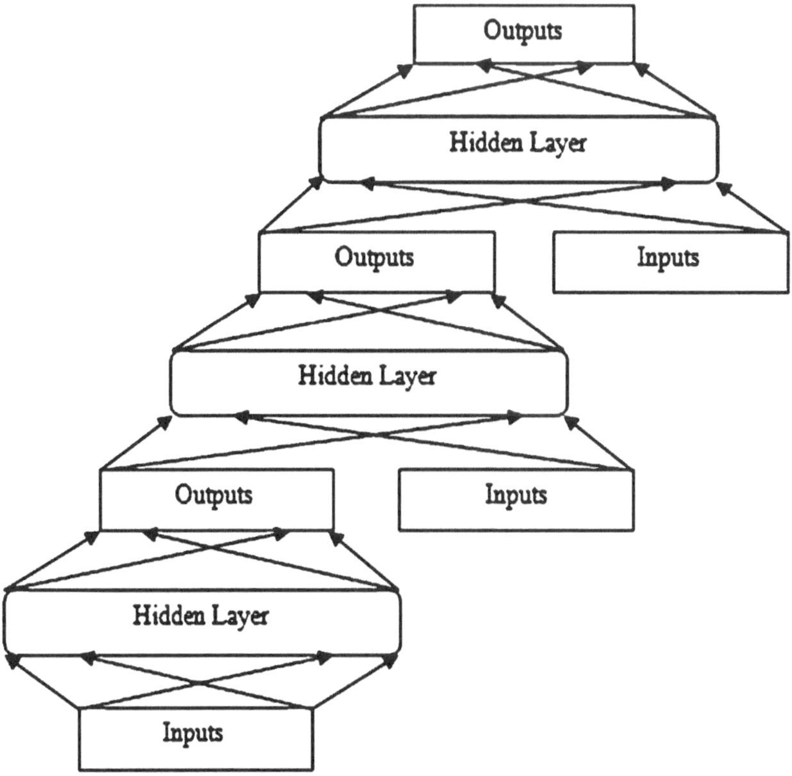

FIGURE 1.13 Deep stack network.

DSN allows the training of independent modules. It works well by providing coaching capabilities. Supervised training is carried out as a distribution for each module, without the entire network penetration.

DSNs can outperform standard DBNs in many problems. This makes DSN popular and efficient network architecture.

1.3 DEEP LEARNING FRAMEWORKS [42,44]

The framework is a platform used to develop software programs and give developers a foundation to build and run their applications. The framework is similar to the application-programming interface (API). It also includes the APIs, along with the technical framework. The framework serves as the basis for programming.

The API provides access to the elements supported by the framework. The framework may include code libraries, compilers, and other programs used in the software development process. DL frameworks provide the building blocks for designing, training, and validating intensive neural networks through interfaces for advanced applications. Good DL frameworks, such as MXNet, PyTorch, TensorFlow, Keras, and so on, are well used in the market.

1.3.1 TENSORFLOW

It is an open-source library of numerical sources using data flow graphs. It has an ever-changing ecosystem of tools, libraries, and public resources. It enables researchers to shrink ML's position in ML. Developers easily build and run ML-enabled applications. This is a product of Google. The main application domains of TensorFlow are as follows:

1. **Image recognition:** Various algorithms are used to classify and identify arbitrary objects and larger images. This is mostly used in engineering applications to identify shapes and analyzing photos on social networks.
2. **Sound recognition:** This is the well-known use case of TensorFlow. With the proper training and testing of data, neural networks are capable of analyzing audio signals. It gives a proper response to the consumer. Voice recognition is mostly used in security, IoT, and so on. Voice search is used by handset manufacturers.
3. **Time series:** Time series algorithms are used to analyze time series data to extract meaningful statistics. It allows predicting an undefined time interval in addition to generate other types of time series.

1.3.2 PYTORCH

It is an open-source learning library that is used to develop and train network-based learning models. It was primarily developed by Facebook's AI research team. It has been used with Python and C++. This is a Python package that provides advanced features as mentioned below:

- Tensor computes with fast GPU acceleration.
- Grade deep neural network is built on an autograde-based system.
- Python packs, NumPy, SciPy, and Cython can be used.

1.3.2.1 APPLICATIONS OF PYTORCH

- **Handwriting recognition:** This includes the understanding of human handwriting and its inconsistency from person to person and in all languages. Facebook's AI CEO, Yann LeCun, founded that CNNs could produce handwritten numbers.
- **Image classification:** PyTorch can be used to create specialized network structures called CNNs. These many CNNs have pictures of something like, say, a kitten, and more like how the human brain works, if CNN sees a kitten image data set, they should be able to boldly point to a new image of the kitten.
- **Text generation:** It helps in generating text which trains an AI model on a specific text and generates an output on what it has learned.

1.3.3 MXNET

It is an intensive learning framework designed for all functionality and flexibility. This allows you to combine the flavor of the symbolic system with the necessary processes of increasing efficiency and productivity. Based on this, there has been a variable dependency schedule that automatically matches the symbolic and critical functions on the fly. Symbolic visualization is the overlay graphical layer that makes fast and efficient memory.

The library is portable and lightweight and weighs on many GPUs and machines. Some existing DL frameworks have drawbacks, and so users need to learn another system for a programming taste. While MXNet solves this problem by providing scalability, portability, and programmability features. MXNet models are portable so they are suitable for small amounts of memory. Therefore, they can train their model in the cloud and ship it. MXNet can also scale multiple GPUs and machines. It supports languages like C++, R, Python, Perl, and so on.

1.3.4 KERAS

It is an open-source neural network library written in Python running on Theano or TensorFlow. It is designed to be modular, fast, and straightforward to use. It was built by Google engineer Francisco Chola. Keras does not perform low-level integration. Instead, it uses another library called "backend" to try it out. Therefore, Ceres is a low-level API for low-level APIs, running on TensorFlow, CNTK, or Theano.

Keras' top-level APIs handle how we build models, define layers, or set multiple input models. At this stage, it complicates the model with loss and optimization, training process with appropriate functionality. It does not move computational graphs like ground APIs. It creates tens or thousands of variables because it is handled by the backend engine.

A tabular representation of the most commonly used frameworks is listed in Table 1.1. Each row in the table corresponds to an open-source framework that is more visible followed by the developer group, supported language, and the appropriate application.

1.4 CONCLUSION

Today, DL has had very successful results in multi-image classification, visual labeling, and human body posture recognition. Intensive teaching architecture works best on primary education. DL networks can be integrated into a cohesive solution using the available open-source framework. DL involves many constructs. These features can establish solutions to various problem areas. These solutions can be extended by existing networks that allow for re-entry of previous entries to be checked ahead of time. It is difficult to construct such intensive learning structures. Many open-source solutions, such as Caffe, DirectLining4J, TensorFlow, and DDL, are available to get everything up and running quickly. The first neural network was made in the year 1986. It has been known as RNN. In this neural network, the input is fed to the many hidden layers. In the hidden layers, the data are passed

TABLE 1.1 List of Deep Learning Framework.

Name	Develop by	Interface	Application domains
TensorFlow	Google	Python, C, C++, Java, R, Go	Image recognition, voice/sound recognition, video detection, text-based applications, time series
PyTorch	Facebook	Python	Handwriting recognition, image classification, text generation (natural language processing)
MXNet	Apache	C++, Python, Julia, Go, JavaScript, MATLAB, R, Scala, Perl, Wolfram Language	Used for scalable enterprise applications, supports flexible programming model in many languages
Keras	Google	Python, R	Prediction, feature extraction, and fine-tuning
Caffe2	Facebook	Python, MATLAB	Image classification and image segmentation, RNNs
Deep Learning 4j (DL4j)	ML Group, San Francisco	Python, Java, Scala, Clojure, and Kotlin	Security applications, recommendation system, parallels, and distribution request
Deeplearn.js	Google	Java Script	For building applications distributed in Hadoop and the Spark framework, an application running in the browser
Microsoft Cognitive Toolkit	Microsoft Research	Python, C++, Command Line	Image, handwriting, and speech recognition problems

from one hidden layer to another hidden layer. Then at last, the data are fed to the output layer. RNN is a less accurate network. In this network, extracted data are passed in single direction only. After RNN, LSTM neural network was developed which consist of feedback layers too. In this, the data are processed between the layers bidirectionally. It consists of three gates that are entry, exit, and forget gates. Image classification was not performed so effectively in this network; therefore, CNN was developed. CNN is capable to classify images effectively and in less time. Later CNN was also found to be used for voice recognition too. It basically consists of two phases that are feature extraction and classification. DBN came after CNN which is based on RBM network. This network contains several RBM networks. Each previous RBM network outputs are fetched as an input for next RBM network. DBN is a collection of various RBM network layers. DSN is like DBN as it also contains different networks in it. In DBN, there was only one network used. While in DSN, hybrid networks are used where each previous network outputs are given as inputs for next network. It is thus considered as a more accurate network than DBN in some applications. GRU network was developed to overcome the limitations of LSTM. It consists of only two gates, that is, refresh and reset gates, due to which the performance of GRU network is considered to be more efficient than LSTM network. Hence, as the time evolved, networks were developed. Each new network was more efficient than the previous network. Many frameworks are developed to implement these networks for different applications and different programming languages. Some commonly used framework is TensorFlow, PyTorch, MXNet, Keras, and Caffe2.

TensorFlow is used for image processing, speech recognition, and text analysis. It is used in programming languages, such as Python, Go, C++, C, Java and so on. PyTorch framework is used in Python programming language. It is used for image classification, text generation, and for many more purposes. MXNet framework was developed by Apache. It is used in many programming languages such as Julia, Go, JavaScript, MATLAB, R, Scala, and Perl. It is supposed to support flexible programming model in many languages. Other frameworks such as Keras, DeepLearning4j, Deeplearn.js, and Microsoft Cognitive Toolkit are used in various networks. All these networks are used for various applications according to their functionality. Hence, to build an efficient and accurate network, an appropriate network and framework are being chosen. Each network and framework have their own advantages and by choosing correct network and framework, an efficient and accurate model can be developed.

KEYWORDS

- **deep learning**
- **machine learning**
- **artificial intelligence**
- **image recognition**
- **speech translation**

REFERENCES

1. https://developer.ibm.com/technologies/artificial-intelligence/articles/cc-machine-learning-deep-learning-architectures/
2. Kalchbrenner, N.; Espeholt, L.; Simonyan, K.; van den Oord, A.; Graves, A.; Kavukcuoglu, K. Neural Machine Interpretation in Linear Times. *arXiv* **2017**. arXiv: 1610.10099v2.
3. Some, K. *The History, Evolution and Growth of Deep Learning*. https://www.analyticsinsight.net/the-history-evolution-and-growth-of-deep-learning/.
4. Marr, B. *A Short History of Deep Learning: Everyone Should Read.* https://www.forbes.com/sites/bernardmarr/2016/03/22/a-short-history-of-deep-learning-everyone-should-read/#39ac3c195561.
5. Pak, M.; Kim, S. A Review of Deep Learning in Image Recognition. In *2017 4th International Conference on Computer Applications and Information Processing Technology (CAIPT)*, 2017, August, IEEE; pp 1–3.
6. Zhou, S. K.; Greenspan, H.; Shen, D., eds. *Deep Learning for Medical Image Analysis*; Academic Press: Cambridge, MA, 2017.
7. O'Shaughnessy, D. Automatic Speech Recognition: History, Methods and Challenges. *Pattern Recognit.* **2008,** *41* (10), 2965–2979.
8. Arora, S. J.; Singh, R. P. Automatic Speech Recognition: A Review. *Int. J. Comput.* **2012,** *60* (9), 34–44.
9. Ma, J.; Sheridan, R. P.; Liaw, A.; Dahl, G. E.; Svetnik, V. Deep Neural Nets as a Method for Quantitative Structure–Activity Relationships. *J. Chem. Inf. Model.* **2015,** *55* (2), 263–274.
10. Zhu, Y.; Ouyang, Q.; Mao, Y. A Deep Convolutional Neural Network Approach to Single-particle Recognition in Cryo-electron Microscopy. *BMC Bioinform.* **2017,** *18* (1), 1–10.
11. McClard, C. K.; Arenkiel, B. R. Neuropeptide Signaling Networks and Brain Circuit Plasticity. *J. Exp. Neurosci.* **2018,** *12*, 1179069518779207.
12. Hinton, C.; Miyamoto, K.; Della-Chiesa, B. Brain Research, Learning and Emotions: Implications for Education Research, Policy and Practice. *Eur. J. Educ.* **2008,** *43* (1), 87–103.
13. Alipanahi, B.; Delong, A.; Weirauch, M. T.; Frey, B. J. Predicting the Sequence Specificities of DNA- and RNA-binding Proteins by Deep Learning. *Nat. Biotechnol.* **2015,** *33* (8), 831–838.
14. Zhou, J.; Theesfeld, C. L.; Yao, K.; Chen, K. M.; Wong, A. K.; Troyanskaya, O. G. Deep Learning Sequence-based Ab Initio Prediction of Variant Effects on Expression and Disease Risk. *Nat. Genet.* **2018,** *50* (8), 1171–1179.
15. Trischler, A.; Ye, Z.; Yuan, X.; Suleman, K. Natural Language Comprehension with the Epireader. *arXiv* **2016**. Preprint arXiv:1606.02270.
16. Nivre, J. Towards a Universal Grammar for Natural Language Processing. In *International Conference on Intelligent Text Processing and Computational Linguistics*; Springer: Cham, 2015, April; pp 3–16.
17. Dong, D.; Wu, H.; He, W.; Yu, D.; Wang, H. Multi-task Learning for Multiple Language Translation. In *Proceedings of the 53rd Annual Meeting of the Association*

for Computational Linguistics and the 7th International Joint Conference on Natural Language Processing (Volume 1: Long Papers), July, 2015; pp 1723–1732.
18. Jiang, S.; Armaly, A.; McMillan, C. Automatically Generating Commit Messages from Diffs Using Neural Machine Translation. In *2017 32nd IEEE/ACM International Conference on Automated Software Engineering (ASE)*, IEEE, 2017, October; pp 135–146.
19. Bahdanau, D.; Cho, K.; Bengio, Y. Neural Machine Translation by Jointly Learning to Align and Translate. *arXiv* **2014**. Preprint arXiv:1409.0473.
20. Wu, R.; Yan, S.; Shan, Y.; Dang, Q.; Sun, G. Deep Image: Scaling Up Image Recognition. *arXiv* **2015,** *7* (8). Preprint arXiv:1501.02876.
21. Lecun, A.; Cortes, C.; Burgess, C. *Handwritten Data for MNIST*, n.d. yann.lecun.com.
22. Malhotra, N. K.; Birks, D. F. *Marketing research: An applied approach.* Pearson Education, 2007.
23. Pawade, D.; Jain, M.; Sarode, G. Methods for Automatic Text Generation. *i-Manager's J. Comput. Sci.* **2016,** *4* (4), 32.
24. Smith, G. W. and Leymarie, F. F. The Machine as Artist: An Introduction. *Arts* **2017,** 6 (2), p. 5.
25. Yao, K.; Peng, B.; Zhang, Y.; Yu, D.; Zweig, G.; Shi, Y. Spoken Language Understanding Using Long Short-term Memory Neural Networks. In *2014 IEEE Spoken Language Technology Workshop (SLT)*, 2014, December, IEEE; pp 189–194.
26. Huang, C. L.; Hori, C.; Kashioka, H. Semantic Inference Based on Neural Probabilistic Language Modeling for Speech indexing. In *2013 IEEE International Conference on Acoustics, Speech and Signal Processing*, IEEE, 2013, May; pp. 8480–8484.
27. Gehring, J.; Auli, M.; Grangier, D.; Yarats, D.; Dauphin, Y. N. Convolutional Sequence to Sequence Learning. *arXiv* **2017**. Preprint arXiv:1705.03122.
28. Miotto, R.; Wang, F.; Wang, S.; Jiang, X.; Dudley, J. T. Deep Learning for Healthcare: Review, Opportunities and Challenges. *Brief. Bioinform.* **2018,** *19* (6), 1236–1246.
29. Ahmed, E., Jones, M. and Marks, T. K. An Improved Deep Learning Architecture for Person Re-identification. In *Proceedings of the IEEE Conference on Computer Vision and Pattern Recognition*, 2015; pp. 3908–3916.
30. Zhang, M.; Cui, Z.; Neumann, M.; Chen, Y. An End-to-end Deep Learning Architecture for Graph Classification. In *Thirty-Second AAAI Conference on Artificial Intelligence*, 2018, April.
31. Jaeger, H. *Tutorial on Training Recurrent Neural Networks, Covering BPPT, RTRL, EKF and the "echo state network" approach.* GMD-Forschungszentrum Informationstechnik: Bonn, 2002, Vol. 5; p 01.
32. https://en.wikipedia.org/wiki/Recurrent_neural_network.
33. https://machinelearningmastery.com/gentle-introduction-long-short-term-memory-networks-experts/.
34. https://en.wikipedia.org/wiki/Long_short-term_memory.
35. https://en.wikipedia.org/wiki/Gated_recurrent_unit.
36. Li, M. Tutorial on Backward Propagation through Time (BPTT) in the Gated Recurrent Unit (GRU) RNN.
37. Shin, H. C.; Roth, H. R.; Gao, M.; Lu, L.; Xu, Z.; Nogues, I.; Yao, J.; Mollura, D.; Summers, R. M. Deep Convolutional Neural Networks for Computer-aided Detection:

CNN Architectures, Dataset Characteristics and Transfer Learning. *IEEE Trans. Med. Imaging* **2016,** *35* (5), 1285–1298.

38. Khan, A.; Sohail, A.; Zahoora, U.; Qureshi, A. S.; A Survey of the Recent Architectures of Deep Convolutional Neural Networks. *Artif. Intell. Rev*. **2020,** 1–62.
39. Hinton, G. E. Deep Belief Networks. *Scholarpedia* **2009,** *4* (5), 5947.
40. https://missinglink.ai/guides/neural-network-concepts/deep-belief-networks-work-applications/.
41. Perwej, D. Y. An Evaluation of Deep Learning Miniature Concerning in Soft Computing. *Int J. Adv. Res. Comput. Commun. Eng*. **2015,** *4* (2), 10–16.
42. https://www.analyticsvidhya.com/blog/2019/03/deep-learning-frameworks-comparison/.
43. https://medium.com/@ODSC/top-7-machine-learning-frameworks-for-2020-7e45164914e1.
44. https://www.simplilearn.com/tutorials/deep-learning-tutorial/deep-learning-frameworks.

CHAPTER 2

Deep Learning in Neural Networks: An Overview

VIDIT SHUKLA and SHILPA CHOUDHARY

Department of Electronics and Communications, G. L. Bajaj Institute of Technology and Management, Greater Noida, India

Corresponding author. E-mail: viditshukla4@gmail.com

ABSTRACT

It was only up till recent times that computer science and its algorithms were sufficient for the application in basic principles. With the expansion in the field of artificial intelligence, the subset Deep learning is impending towards substantial research and advances, creating diverse opportunities. We cannot consider deep learning to be an individual approach; it is rather a collective term which comprises fields from contrasting backgrounds to be associated with the common spine—Deep learning. Basis for the strong approach in deep learning lies in cognizance of the architecture of deep learning. The implementations can be performed vastly in any fields through implication of not just one but numerous algorithms and achieving our goal. The architecture of deep learning has enhanced in the previous years exponentially, and as per demand, the refinement of deep learning implying that the architecture is dynamic.

A few of the most improvised architectures are mentioned below:

- Recurrent neural networks (RNNs)
- Long short-term memory (LSTM)/gated recurrent unit (GRU)
- Convolutional neural networks (CNNs)

Deep Learning in Visual Computing and Signal Processing. Krishna Kant Singh, Vibhav Kumar Sachan, Akansha Singh & Sanjeevikumar Padmanaban (Eds.)

- Deep belief networks (DBN) and deep stacking networks (DSNs)
- Open source software options for deep learning.

The area of implementation for deep learning in problem solving is vast. Feed forward networks are very effective as well as recurrent networks can be a good source for the solution of the deep learning problems. The Framework for deep learning can be implemented in software packages for the useful creation of neural network. The framework needs an implementation on a standardized scale and hence needs industrial experts for the framework to be implemented. The entire framework is in simple terms based on the Diagnosis of the problem and further, evaluating the problem. It is evident that the architecture and framework of deep learning is vast and expanding its horizons to every field possible for implementation.

Therefore deep learning architecture and framework would be vitalized, with step by step conception. The architecture would be simplified as well as illustrated. All the aforesaid architecture like Recurrent neural network, Long short term memory/gated recurrent unit, convolutional, Deep belief—deep stack as well as open source would be simplified as well as illustrated.

2.1 INTRODUCTION TO DEEP LEARNING

"Deep learning is the new crude oil" says the *Economic Times*. The said proposal is now manifesting very well in each aspect of technological advancements in fields of artificial intelligence associated with machine learning and deep learning, but not limited to.[1]

Understanding deep learning in layman terms is related to story of the availability of the fossil fuels that are present beneath the Earth's surface. The mere knowledge of presence of availability of fossil fuels is not helpful. The method of extraction, its purification, and then maximum use for advantage is must. Similarly, to get anything worthwhile from data, deep learning is a very useful tool to detect and extract the required data for beneficial purposes. The only difference is that unlike crude oil, we will never be in shortage of data.

Before directly diving into deep learning, we must first know the difference between the three leading and well-known methods—artificial intelligence, machine learning, and deep learning. The three are as close to each other as they are different. Artificial intelligence is a broad term.

Subset to it is machine learning. Further, subset of machine learning is deep learning.

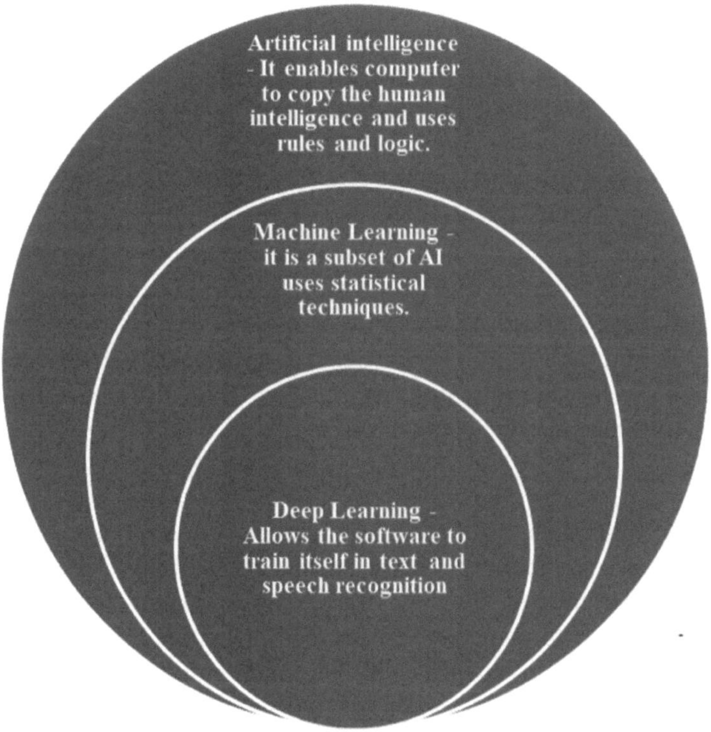

FIGURE 2.1 Artificial intelligence, machine learning, and deep learning.

The broadest of the three terms is artificial intelligence. It includes all the techniques and methods that can be implemented to learn from human actions and copy them. On the other hand, machine learning is a subset which works on a feedback nature, to learn from its own actions, and improve the mechanism, time, as well as actions. Deep learning is clearly the algorithm that is used for the purpose of recognition of image, voice, and to train the software to help them learn themselves. This makes clear about the actions of the three mechanisms.

"Deep learning is a type of the machine learning algorithms that is a high end method to extract as much as data and patterns possible in comparatively less time, through use of dense layers and get features that are of greater use from the raw input data provided."

The modern deep learning models are a result of common neural networks that are used in variable layers. Neural network plays a major role being an integral part of deep learning layers. It is the basis for creation of the layers in the method for deep learning. Each and every process of deep learning is a step to make the data abstract than the last step.[1]

Example—If raw data of a face are provided to the system, the small steps in image recognition is done one by one: in the first step, understand the pixels' arrangement, next the corners and edges, followed by the parts of the face further till it is completely recognized by the algorithm as a face.

Deep learning is all about the increase in the layers of the mechanism. The increased layers are used to transform data. It follows a CAP, which is credit assignment path. It is a series or chain like, which are transformations from input to output. There is depth as a parameter which is based on the factor whether the network is feedforward or feedback. If it is a simple feedforward network, the depth is given as:

$$\text{Depth} = \text{Number of hidden layers} + 1$$

In case of a feedback or recurring network, the depth increases vastly and is potentially unlimited, because of the signal traveling more than once. It is clear that deep learning requires more and more layers for useful and improved performance. The increased layers are a better source for refinement in the raw data provided. The smaller divisions are an integral part of this method.

Deep learning was coined to the machine learning researchers by Rina Dechter in the year 1986, and then later on to artificial neural networks (ANNs) by Igor Aizenberg and colleagues in the year 2000, in the context of Boolean threshold neurons. Although it interesting to find out that very early, authors and researchers Alexey Ivakhnenko and Lapa in 1967 in one of their papers mentioned about supervised, forward, multilayered, perceptron as well as feedback layers were discussed in the chapter.

The earliest deep learning requirement aroused due to incapacity of the machine to understand and read handwritten numbers and words. There was specially built computer vision to help the machine solve the problem of understanding the handwritten numbers and alphabets by Kunihiko Fukushima in 1980. This was for the postal services to segregate the zip codes in an easier way. But these early softwares had downsides, like there was a lot of requirement for training before operation.

Later, 2D and 3D objects became a matter of concern to be understood and identified by the machine. The neural networks had become an obsolete and difficult method for the process and, therefore, Gabor filters and support vector machines were later used for the same task in a lower computational cost. ANN along with its components of shallow learning and the other deep learning have been tried to understand, implement, and learn since a lot of years. These methods never outperformed nonuniform internal-handcrafting Gaussian mixture model/Hidden Markov model technology implied on generative models of speech trained discriminatively.[2]

The deep learning mechanism learned much by the neural networks speech recognition system, which was known as long short-term memory network (LSTM); this is a type of neural network recurrent in nature, which was proposed and introduced by Hochreiter and Schmidhuber in early 1997. LSTM recurring neural networks solved the disappearing gradient issues can be resolved by "very deep learning" tasks that need memories of events that happened thousands of discrete time steps earlier, they are vital for speech recognition to be carried out.[3]

This requires memory for the use of all the prior steps, which are required for speech. LSTM gained much popularity in 2003, after which it grew to be a better version of the network.[4] The same was implemented by Google in 2001, which proved to be a boon for the speech recognition system for Google, Google voice search.

Deep learning is a flagship method that deals with computer vision and automatic speech recognition (ASR) in a different way. Solution is provided by the help of evaluation sets such as TIMIT (ASR) and MNIST (image classification), and a range of large-vocabulary speech recognition tasks have continuously improved in a steep manner. Convolutional neural networks (CNNs) were superseded for ASR by CTC for LSTM, although talking about their success rate, it is higher in computer vision. With time, the developments increased and the deep leaning and neural networks gained higher preference for speech recognition, after Google improvised the method in speech recognition. This led to hardware association of deep learning as well. More advancement in hardware has made the deep learning process important.

Nvidia did a great work in the year 2009. It made reasonable advances in the deep learning implementation methodologies. It revolutionarized the idea of deep learning all at once. Nvidia named it “big bang” where it associated its graphical processing unit (GPU) and trained them for special uses. This was a sheer advance move by Nvidia to implement deep learning more and more.

Following the same, Google used the Nvidia GPU in "Google Brain" and made its use for the creation of DNN. GPU are very well versed for the data in form of matrix and fragmented basic forms. The deep learning process is in fact the one which considers the data in matrix form. This step created harmonious work of deep learning with hardware GPU. They reduce the time required, that is, running time and if further used in a specialized manner, can be very suitable for deep learning.

This was not just when things moved toward revolution. In the year 2012, George E. Dahl and his team won the "Merck Molecular Activity Challenge" with the implementation of multitask deep neural networks to predict the biomolecular target of one drug. Later in the year 2014, Hochreiter's team implemented deep learning to detect off-target and toxic effects of environmental chemicals in nutrients, household products, and drugs and subsequently won the "Tox21 Data Challenge" of NIH, FDA, and NCATS (ToX data challenge results).

This was when the use of deep learning unfolded in a critical way.

In the year 2011 and 2012, a lot of papers, thesis, and research went in direction of deep learning. Backpropagation, feedback, and feedforward networks were already known but there was need for better use of the implementations for neural networks. Faster implementations and recognitions were increasing. Image classification did not stop where it was, instead it kept on increasing from the mere recognition of plain words to their definition, description, and detailing. This was made with the help of a combination of CNNs and LSTMs.

It is evident from here that deep learning has always been dependent on neural networking and layers, which were constantly on improvement mode for the creation of better, fast, and accurate deep learning is a very crucial aspect in increasing the efficiency of deep learning. Greater advancements in neural networking clearly build an improved mechanism.

2.1.1 ARTIFICIAL NEURAL NETWORK

For further analysis into deep learning, we will have to know about deep learning first. The neural network is connectionist system. The ANN is completely inspired and made with the base of the biological neural network. The inspiration is from the human brain which itself a very big neural network. These complex networks exist in our brain. The constant improvement and learning is the essential for the neural network to grow and also refine its own results.

Example—In the process of image recognition, if there are several images of flowers, the neural network learns the structure of flowers and hence labels or saves the structure as flower, and hence whenever any image of a flower comes, it is automatically displayed by the system as flower.

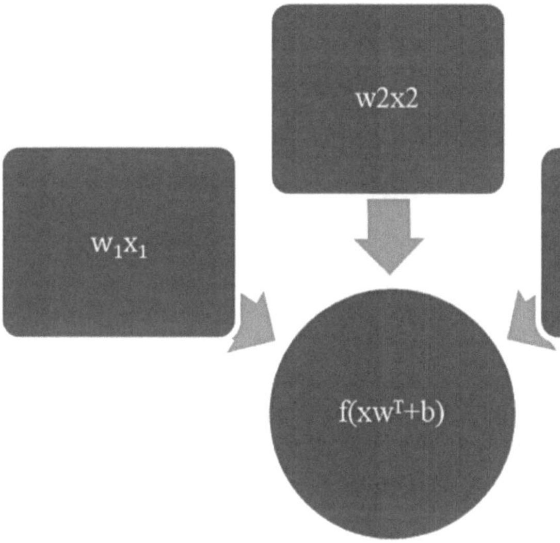

FIGURE 2.2 A simple artificial neural network.

ANN is a combination of small cell known as neurons. They have their own weights that change, as per the learning. The initial layer is the input layer. The last layer is the output layer. There are several layers in between these two, which are variable as per the learning. The use of neural network has increased vastly in fields of image, speech, voice recognition, and assessment.

2.1.2 DEEP NEURAL NETWORK

If there are multiple layers in between input and output layers, it is known as deep neural network. It is important because we get the correct relation to the input–output. It works on the principle of probability.

Example—If the input provided is of a house, the neural network layers will pass on to find the percentage of the relation of the image inserted with its resemblance with house. Now, here if the percentage is above the minimum limit, the output is given as house.

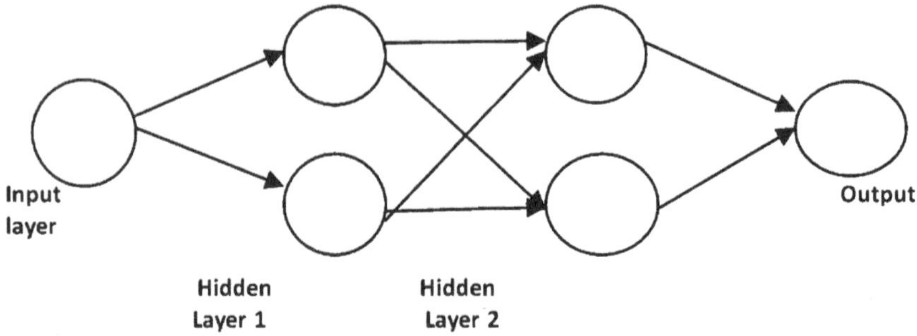

FIGURE 2.3 Hidden layer representation.

2.1.3 PROBLEMS FACED

The neural networks and deep neural networks face issues are somewhat on similar basis and arise due to the lack of trainings.

The major two problems faced are as follows:

1. **Computational time:** It is evident that in the case of multilayers the data flow is delayed and due to large number of layers in between the input and output. The delay in the computational time is due to the numerous layers in between. There are methods to reduce the slowdown.
   Example—Batching is the technique used to reduce the computational time.
2. **Overfitting**—Overfitting is the implementation of models or processes, which violate Occam's razor. Taking an example by inclusion of extraflexible parameters that are ultimately optimal, otherwise with the implementation of more difficult than is ultimately optimal.
   Example—In the presence of numerous adjustable parameters, if we get a function μ which is a linear function of the parameters, replacement of the numerous parameters take place, and hence, it becomes simple for the neural network and hence speed increases.

There are endless methods to still be discovered for the better usage of speed reduction techniques to help reducing time for the computation.

2.1.4 APPLICATIONS

2.1.4.1 IMAGE RECOGNITION

It is very well generalized and known that the image recognition has been implementing deep learning methodologies and implementing them for handwriting recognition, areas of signature verifications, and other implementations.

The image recognition system has clearly advanced the human level of image recognition and is much accurate than humans.

Example—Facial dysmorphology novel analysis.

2.1.4.2 AUTOMATIC SPEECH RECOGNITION

Very deep learning neural networks were implemented on a large scale. LSTM recurrent neural networks (RNNs) were implemented at that time for the learning of the network. The network learns and holds a lot of data. The speech recognition system was built for small-scale initially; later on, the network was increased and made deep. Hence, the networks became deep neural networks and the speech recognition became better and better.

2.1.4.3 TOXICOLOGY

The neural networks have emerged to be pioneers in toxicology. It works on the basis of target and nontarget effects. The minimum required target and nontarget percentage is calculated. The drugs are approved by the board only after the required criteria are met.

2.1.4.4 RELATIONSHIP MANAGEMENT

The marketing techniques require learning to target the correct audience for the organization. It helps in the organization to find the profits per customer as well which is beneficial for the organization.[5]

2.1.4.5 MEDICAL IMAGE ANALYSIS

In the field of medical analysis, the deep learning process gives result and studies the cell classifications, image enhancement, and organ sedimentation as well. This is used with the help of comparative study.[6]

2.1.4.6 ADVERTISEMENT

The target audience has been an issue for the media sector, which is being resolved very easily by the deep learning neural networks. The target audience is filtered out by deep neural networks and the advertising media focuses on selling their brands to the target audience. This significantly improves the efficiency of the advertisement company and helps them to choose wisely.

2.1.5 RELATIONS OF THE BIOLOGICAL NEURAL NETWORK WITH SELF-LEARNING

In 1990, the cognitive neuroscientists suggested that the basic development of the brain relates to the process similar to learning's in the deep networks. As the information passes through the layers, self-learning starts and thus the network learns in a similar way the human brain learns since childhood. Every learning is inspired from the brain itself.[7]

Example—The calculations done by deep learning units have a chance to be same as to those of actual neurons and neural populations. In the same manner, the results developed by deep learning models are comparatively more as measured in the primate visual system, both at the single-unit and at higher levels.[8–10]

2.1.5.1 WELL-KNOWN COMMERCIAL PROJECTS

Two of the best-known commercial forms are from Google—DeepMind Technology. It had successfully played a video game with just pixels as input. The other well-known form was Facebook's artificial intelligence mechanism, which performs automatic tagging of the photographs uploaded. This is pure use of very deep neural networks.[10]

Some smaller commercialization has been done like the University of Austin at Texas did create a new platform, which was unique. It helped the robots and computers to learn with the help of a human, which provides instructions.[11]

There have also been other uses of the same.

2.2 EVOLUTION OF DEEP LEARNING ARCHITECTURE

TABLE 2.1 Evolution of Deep Learning Architecture through Years.

Years	Architecture
1990–1995	RNN
1995–2000	CNN
2005–2010	DBN
2010–2015	DSN
2010–2015	GRU

The 20 years of development in the algorithm is shown in Figure 2.3. Deep learning neural networks have evolved a lot through its own learning and development.[12] The RNN is clearly made in the early 1990s and is still the most efficient and widely used form.[13]

TABLE 2.2 Different Architectures and Their Applications.

Architecture	Application
RNN	Speech recognition handwriting recognition
LSTM/ GRU	Image captioning, text compression, handwriting recognition, gesture recognition, natural language text compression, speech recognition
CNN	Image recognition, natural language processing, video analysis
DBN	Image recognition, failure recognition, video analysis
DSN	Information retrieval, continuous speech recognition

2.3 ARCHITECTURE OF DEEP LEARNING

2.3.1 RECURRENT NEURAL NETWORK

The RNN is the oldest network architecture and is the most used applied network.[14] The conversion of nonsequential to sequence is the main usage of RNN. The model sequence can have variable lengths. The weights are shared across time steps.[15]

They are made from the two major networking methods, feedforward and back-propagation methods.[16]

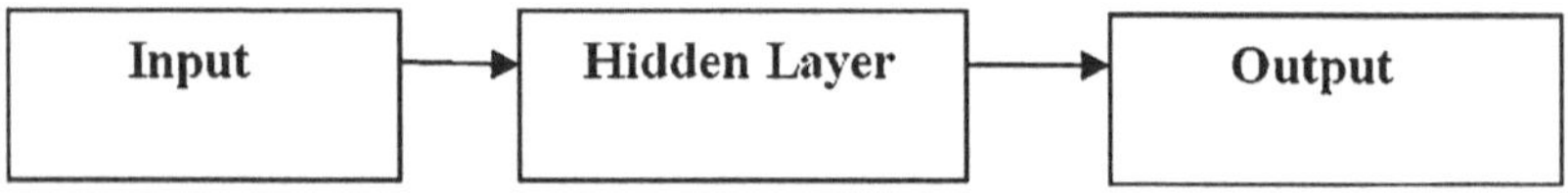

FIGURE 2.4 Basic flow diagram of RNN.

2.3.1.1 FORMULAS FOR RRN

Formula for calculating current state

$$\hbar_t = f(\hbar_{t-1}, \mu_t)$$

where $\hbar_t$ represents current state; $\hbar_{t-1}$ represents previous state; and μ_t represents input state.

Formula for tan h, which is activation function

$$\alpha_t = W_{\varepsilon y} \varepsilon_t$$

where $W_{\varepsilon y}$ represents weight at recurrent neuron and ε_t represents weight at input neuron.[17]

2.3.1.2 BENEFITS OF RECURRENT NEURAL NETWORK

1. RNN is very well capable of remembering and storing the values over time. It stores all the values and does not forget. This opens up a big advantage for the time prediction problems that can benefit from it. It is later discussed as LSTM.[16]
2. Pixel neighborhood is a factor which gets boosted up with the help of RNN when they are implied with convolution layers.[18]
3. RNN needs extensive and tiresome efforts for training.

2.3.1.3 PROBLEMS RELATED TO RECURRENT NEURAL NETWORK

1. Gradient fades away which creates a problem. Also, exploding problems create an issue.[19]

2. RNN is not so easy for the user to train as per the input–output scenario.[20,21]
3. If the length of the sequence is very big, the tan *h* cannot be used as the activation function for such long sequences, thus creating a problem.[22]

The RNN is a different type of network which is artificial in nature. The objective of it is to maintain an internal state. For the same, it adds cycles in the graph of the network to achieve it.[23]

The topics like LSTM, GRU, and NTM are used for the architecture of the deep learning as the base of the future advance topics which need to be studied and improvised.[24]

- RNN or recurrent neural networks—this RNN includes three classifications which are as follows:
  1. Neural history compressor
  2. Recursive neural networks also known as RNN
  3. Fully recurrent networks
- Neural Turing machines
- Gated recurrent unit neural networks
- LSTM

2.3.1.4 FULLY RECURRENT NETWORK

The fully RNNs are the ones which have weighted connection to every other element with a single feedback.

Also, every synapse has a modification enabled real value weights. For supervised learning, real-value sequenced vectors are given in the input, one by one. In case of reinforcement learning, no target signals are provided. The correct ones are rewarded with give the network learning.[25]

2.3.1.5 RECURSIVE NEURAL NETWORK

The RNN is just similar but different form of recursive network. Recursion enhances the process of refined results by the use of weights again and again and giving fine results from the processing also after training.[26]

Training for this is imparted provided with the help of gradient descent method by subgradient methodology implementation.[27]

2.4 LONG SHORT-TERM MEMORY NETWORK

The LSTM is a type of architecture, artificial in nature of the RNN. It has great importance in the field of deep leaning applications. It varies with feedforward networks in the ground that feedforward networks do not have feedback provision. The LSTM has feedback system.[28] It is capable of processing multiple data points, which are video and audio signals, rather than just the single data point. Major uses include handwriting recognition, predictions, network traffic detection, and voice recognition. It includes a **cell,** an **input gate,** an **output gate,** and a **forget gate.**[29]

The LSTM was proposed and introduced to the world by Sepp Hochreiter and Jürgen Schmidhuber in the year 1997 for the solution of the gradient problem. The initial model was designed without forget gate. The forget gate was later introduced in the year 1999.[30] It is also known as the Keep gate. It allows the LSTM for resetting itself. Similarly later on, Peephole Connections were added by Gers and Schmidhuber and Cummins. With time, many reforms of the network were introduced.[31]

Even in recent years, Tech Giants like Google, Apple, and Microsoft are using the LSTMs as the basic component even in their new products. Taking an example, Google used LSTM technology for speech recognition on the smartphone. Others like Alexa and Siri also implement the same.[32]

The basic idea for LSTM is that they hold a record of arbitrary long-term dependencies in the input sequences. The issue raised of vanilla RNN is of a great concern and is ever existing. The phenomenon of vanishing has always taken place to create hurdle in the way. The same was tried to be solved in many ways. The solution that came up was the use of LSTM as an implementation for the reduction of this problem hence eliminating it.[33]

2.4.1 ARCHITECTURE OF LSTM

The most common architecture of the LSTM is shown in Figure 2.5. It consist of the following parts:

- Cell (the memory of LSTM)
- Forget gate
- Input gate
- Output gate

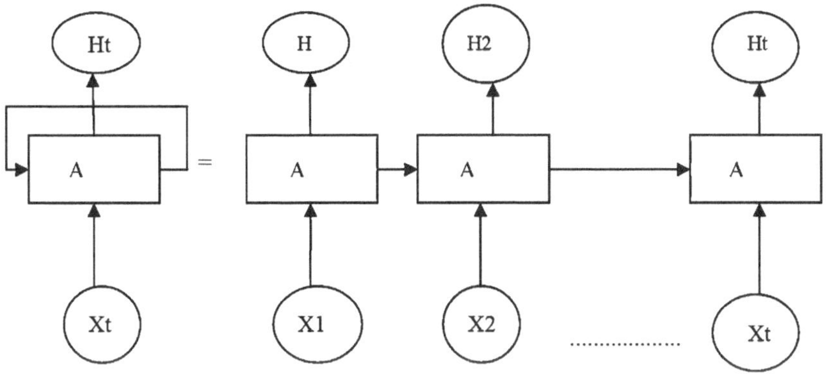

FIGURE 2.5 LSTM architecture.

Here, the cell is the one responsible for the control, tracing, and keeping records of elements in the input sequence. Input gate alters the flow of input, the rate of flow of new data, and the output regulates the level as up to which the value in the cell is taken into consideration when we perform calculations to find out the output for LSTM. Logistic sigmoid function is the activation function of the LSTM gates. A few of the relations between are recurrent. It is to be noted that the learning plays an important role for the gates to open.

2.4.2 VARIATIONS IN LSTM NETWORK

The LSTM network usually varies in some or the other ways as created by the user. Here, three variations found in LSTM are mentioned and explained.

1. Gers and Schmidhuber variation—Also known as peephole LSTM, this was the variation introduced in the year 2000. Gers and Schmidhuber added a peephole connection. The gate layer has an overview of cell state.[34]

Formulation is given below:

2. The peephole convolution LSTM—This system is denoted by* which is same as of convolution symbol.

$$F_t = \sigma_g(W_f x_t + U_f c_{t-1} + b_f)$$
$$I_t = \sigma_g(W_i x_t + U_f c_{t-1} + b_i)$$

$$o_t = \sigma_g (W_o x_t + U_o c_{t-1} + b_o)$$
$$c_t = f_t o c_{t-1} + i_{to} \sigma_c (W_c x_t + b_c)$$
$$h_t = \sigma_h (o_t o c_t)$$

The formula is given below:

$$F_t = \sigma_g (w_f * x_t + U_f * h_{t-1} + V_f o c_{t-1} + b_f)$$
$$I_t = \sigma_g (w_i * x_t + U_i * h_{t-1} + V_i o c_{t-1} + b_i)$$
$$C_t = f_t o c_{t-1} + i_t o \sigma_c (W_c * x_t + U_c * h_{t-1} + b_c)$$
$$O_t = \sigma_g (W_o * x_t + U_o * h_{t-1} + V_o O c_t + b_o$$
$$h_t = \sigma_t (o_t o c_t)$$

2.5 CONVOLUTIONAL NEURAL NETWORK

The CNN is a different and specially trained kind of neural network model made for working with 2D image data. But they can be improvised with 1D and 2D data as well. This kind of network is named as CNN after one of its layers is of convolution mechanism. It is multilayered, which is derived from the base of animal visual cortex. Its basic usage is for visual imagery. They are as well called by as **shift invariant** or **space invariant artificial neural networks (SIANN),** which is on the basis of shared-weights architecture and translation invariance characteristics.[35]

The first CNN was designed by Yann LeCun; in the times when the architecture paid attention on handwritten character recognition, taking an example of postal code interpretation.[36]

CNNs imply comparative from others, less preprocessing with comparison to other image classification algorithms. The neural network is designed such as to learn and memorize the way of representations of the handwritten notes. This freeness from the previous knowledge and human engineering in feature design is a very big benefit.

2.5.1 ARCHITECTURE OF CNN

1. The LeNet CNN architecture is composed of numerous layers, which are applied feature extraction, then further classification. It dates back to 1998.[37]

The image is bifurcated into receptive fields which save into a convolutional layer, following which it then extracts features from the input image. This is also a seven-layer architecture. The final output layer of this network is a set of nodes that identify features of the image (in this case, a node per identified number). The network is trained using backpropagation.

2. AlexNet is another architecture of CNN and was introduced in the year 2012. It was introduced and developed by SuperVision group. It was quite indistinguishable to the LeNet architecture. AlexNet is the one which does not used tan *h* and implement new ReLu.
3. Google Net was the one introduced in the year 2014. It was as the name suggests, introduced by Google. It was a high accuracy and executed with accuracy as high as 92–93% making error less than 7–8%. It received a lot of awards for the high level of accuracy.
4. VGGNet was the architecture that came into limelight in the year 2014. It uses convolutions that are 3 × 3 in nature. The real matter of concern is that VGGNet uses 138 million parameters that are very much higher any other architecture.[38]

Advancements have resulted in the LSTM to train CNN in such a way to easily break the image and video to be understood.

2.5.2 CLASSIFICATION OF CONVOLUTION NEURAL NETWORK

Invariance to local translation can be a helpful property if we focus more regarding whether if a few of the features are placed in the same manner or not. Taking an example, at the point of finding if an image contains a face, it is not as important for the determination of whether the eyes are there or not. Instead, a slightly estimated method would also be very helpful just to check it is present or not.

2.5.2.1 CONVOLUTION IN COMPUTER SYSTEM

Convolution has been used in the computer vision since very long. The idea is not a fresh one to implement convolution to image data. Earlier, filters were made by the experts in the field of vision of computers, which were later applied to an image to result in a feature map or output from using the filter then make them more efficient in all possible ways.

Taking an example, a 3×3 element filter for detecting vertical lines which is handcrafter is shown as the following:

0.0,	1.0,	0.0
0.0,	1.0,	0.0
0.0,	1.0,	0.0

Using this filter to an image will give a result of in a feature map that only contains vertical lines. Therefore, the implementation of it is as a vertical line detector. Similarly, horizontal line detectors are also created and applied to image. A set of a few, ranging from the value of tens to hundreds and thousands are used and also applied for the same purpose.

Such networks are trained in such a way that they differentiate between the basic compositions of two dissimilar objects.

Example—The network is trained for identifying the features that are useful to differentiate between a cycle and a motorcycle.

Power of Learned Filters

- Multiple filters
- Multiple channels
- Multiple layers

Convolutional layers are the ones which are not given as input to the network.

Taking example—unprocessed pixel values. The other feature of it is that they can be applied to the gates of other output gates as well.

The collection of convolutional layers gives a permit of a hierarchical decomposition of the input.

Examples of convolutional layers—The convolutional layer has two clear parts, 1D and 2D convolution layers.

2.6 DEEP BELIEF NETWORK

DBN network or deep belief networks are the ones that are made up of novel training algorithms. It is a multilayer and has numerous hidden layers. Also, note that the pairs that are formed in DBM are of Boltzmann machine, but of restricted nature. The raw sensory inputs are represented as input layers. The output layer however is treated in a different way. This consist of two levels of learning, supervised fine-tuning, and unsupervised pretraining.[39]

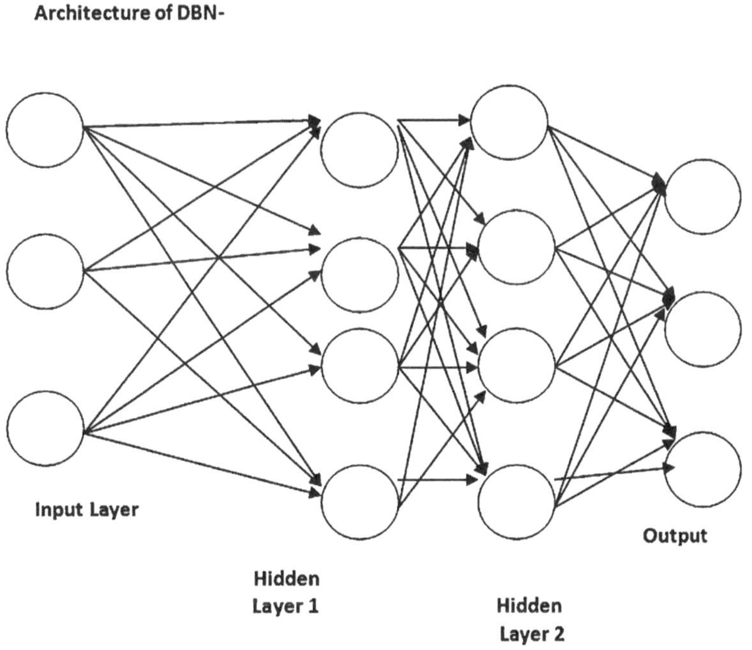

FIGURE 2.6 Architecture of deep belief network.

2.7 DEEP STACKING NETWORKS

Deep convex network are a bit different than that of general deep learning architecture. They are instead a set of individual deep networks. This architecture is an output to one of the difficulties faced with deep learning the most—the topic of complexity training. The deep stacking network is the one which are much advanced than the other. They use the method of parallel computation for the processing in a higher and faster way. There is a high level of efficiency which is due to the linear characteristic by the output. In the very introduction designs the hidden layers are represented as sigmoidal function. The large CPU processors are used by the parallel computations in a very good manner by the use of the DSN.

A few problems faced by the DSN are too much human help and jumping in, slow convergence rate. Also, local minima are issues for the DSN.

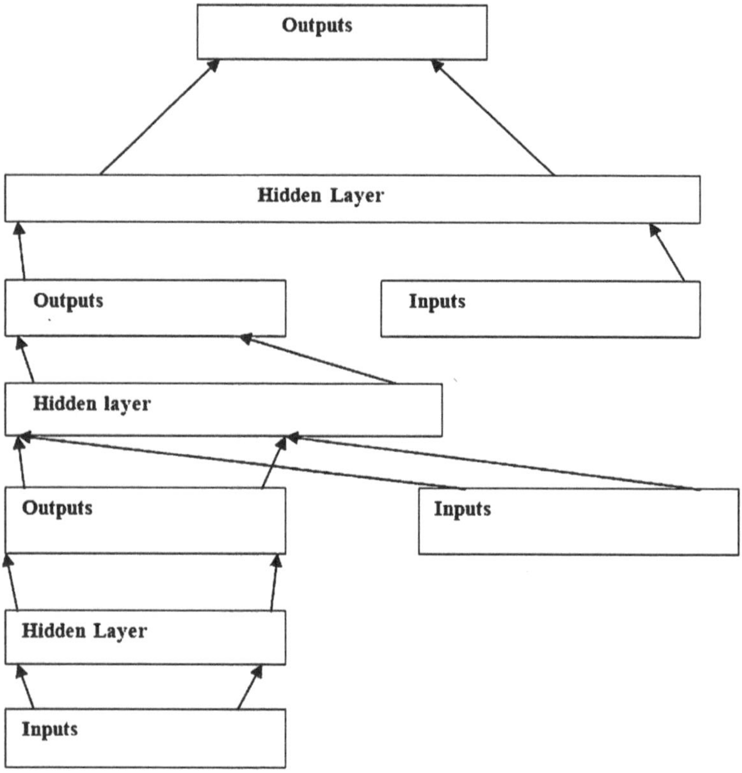

FIGURE 2.7 Architecture of DSN.

2.8 DEEP LEARNING FRAMEWORK

The most commonly used frameworks that are important to be noticed, are listed below:

1. Tensor flow—It is till now, the most well-known deep learning framework. This is because a lot of tech giants. It is a famous framework developed by Google itself.

   It works with Python as a programming language; the other languages like C, C++, and Java, and so on, are just under trial and are not much convenient to be used with tensor flow.

It is capable of running on both the major platforms Android and iOS.

With the helpful tool comes great coding responsibility. Tensor flow not only just gives the helpful results in the deep learning mechanism, but also makes the entire process less difficult. It operates with a **static computation graph**. This implies that initially there is a need to define the graph. Later, process the calculations and, if it is found that there is a need to make favorable editing to the architecture, the model is retrained.

Implementation uses—Handling and experimentation with deep learning and its architecture. Data integration is a part of it, like Graphs, SQL, and so on. Also, it provides a sense of reliance because of being an offering from Google.

2. PyTorch—It was developed specially for Facebook deep learning algorithms. A major difference between the Tensor Flow and PyTorch library is that the PYTorch works with a **dynamically updated graph.** This implies that it permits editing and to the architecture in the method. Also, debuggers like PyCharm can be implied for further uses.

   Implementation uses—It is useful as it has simple and easy training algorithms for the user. It has a lot of trained models already for instant use. It suits for personal scale projects and making prototypes.
3. Sonnet—It is the framework that was created for making complex algorithms. Its objective was very clear relating to the creation of complex algorithms. It was created by DeepMind organization.

   Implementation uses—It consist of high-level object libraries which comprises abstraction in the process of development of the neural networks. The purpose is to make mainly python objects in respect to a target part of the neural network. The achievement that we have from the Sonnet is regarding the uses in research work provided by DeepMind.
4. Keras—It is a relatively new platform for the AI as well as deep learning concerns. It is very well suited for the ones that are well versed with the deep learning algorithms and approaches. Also, to add, it is the highest level of simpleton approach to using TensorFlow, CNTK, or Theano is the high-level Keras shell.

   Implementation uses—It is the most preferred framework as it is very friendly with the user who have initiated their practice. It makes the practice of working and learning properly implied. It is very well designed for new applications because of API.

5. MXNet—This framework is capable of being used on a very large range of devices. This MXNet can be used in a lot of languages like C++, Javascript, R, Julia, and others. The focus is that it is effective on parallel supporting usage of GPUs.
   Implementation uses—The quick problem-solving ability is a key point for the framework. The codework is comparatively easy and can be done by new learners. And, the multiple support of GPUs for processing and work plays a major role in the computational optimizations.

2.9 CONCLUSION

Knowledge of deep learning, its architecture, and framework is important for scholars from any technical background.

The area of implementation for deep learning in problem solving is vast. Feedforward networks are very effective as well as recurrent networks can be a good source for the solution of the deep learning problems. The framework for deep learning can be implemented in software packages for the useful creation of neural network. The framework needs an implementation on a standardized scale and hence needs industrial experts for the framework to be implemented. The entire framework is in simple terms based on the diagnosis of the problem and further evaluating the problem. It is evident that the architecture and framework of deep learning is vast and expanding its horizons to every field possible for implementations.

KEYWORDS

- **framework**
- **architecture**
- **neural networks**
- **artificial intelligence**
- **networks**

REFERENCES

1. *Dhar, V. Data Science and Prediction. Commun. ACM* ***2013,*** *56 (12), 64–73.* doi: 10.1145/2500499 (archived from the original on November 9, 2014, retrieved September 2, 2015).
2. Baker, J.; Deng, L; Glass, J.; Khudanpur, S.; Lee, C.-H.; Morgan, N.; O'Shaughnessy, D. Research Developments and Directions in Speech Recognition and Understanding, Part 1. *IEEE Signal Process. Mag.* **2005,** *26* (3), 75–80.
3. Schmidhuber, J. Deep Learning in Neural Networks: An Overview. *Neural Netw.* **2015,** *61*, 85–117. arXiv:1404.7828. doi:10.1016/j.neunet.2014.09.003. PMID 25462637.
4. Graves, A.; Eck, D.; Beringer, N.; Schmidhuber, J. Biologically Plausible Speech Recognition with LSTM Neural Nets. In *1st Int. Workshop on Biologically Inspired Approaches to Advanced Information Technology, Bio-ADIT 2004*, Lausanne, Switzerland, 2003; pp 175–184.
5. Tkachenko, Y. Autonomous CRM Control Via CLV Approximation with Deep Reinforcement Learning in Discrete and Continuous Action Space, April 8, 2015. arXiv:1504.01840.
6. Litjens, G.; Kooi, T.; Bejnordi, B. E.; Setio, A. A. A.; Ciompi, F.; Ghafoorian, M.; van der Laak, J. A. W. M.; van Ginneken, B.; Sánchez, C. I. A Survey on Deep Learning in Medical Image Analysis. *Med. Image Anal.* **2017,** *42*, 60–88. arXiv:1702.05747. Bibcode: 2017arXiv170205747L. doi:10.1016/j.media.2017.07.005. PMID 28778026.
7. Quartz, S. R.; Sejnowski, T. J. The Neural Basis of Cognitive Development: A Constructivist Manifesto. *Behav. Brain Sci.* **1997,** *20* (4), 537–556. CiteSeerX 10.1.1.41.7854.
8. Yamins, Daniel. L. K.; DiCarlo, J. Using Goal-driven Deep Learning Models to Understand Sensory Cortex. *Nat. Neurosci.* **2016,** *19* (3), 356–365. doi:10.1038/nn.4244. ISSN: 1546-1726. PMID: 26906502.
9. Zorzi, M.; Testolin, A. An Emergentist Perspective on the Origin of Number Sense. *Philos. Trans. R. Soc. B* **2018,** *373* (1740), 20170043. doi:10.1098/rstb.2017.0043. ISSN: 0962-8436. PMC: 5784047. PMID: 29292348.
10. Güçlü, U.; Van Gerven, M. A. J. Deep Neural Networks Reveal a Gradient in the Complexity of Neural Representations across the Ventral Stream. *J. Neurosci.* **2015,** *35* (27), 10005–10014. arXiv:1411.6422. doi:10.1523/jneurosci.5023-14.2015. PMC 6605414. PMID 26157000.
11. Leek, J. "The Key Word in Data Science" Is Not Data, It Is Science. *Simply Stat.* December 12, 2013 (archived from the original on January 2, 2014; retrieved January 1, 2014).
12. Donoho, D. *50 Years of Data Science*, September 18, 2015. Retrieved April 2, 2020.
13. Hayashi, C. What Is Data Science? Fundamental Concepts and a Heuristic Example. In *Data Science, Classification, and Related Methods*; Hayashi, C., Yajima, K., Bock, H.-H., Ohsumi, N., Tanaka, Y.; Baba, Y., Eds.; *Studies in Classification, Data Analysis, and Knowledge Organization*; Springer Japan, 1998; pp. 40–51. doi:10.1007/978-4-431-65950-1_3. ISBN: 9784431702085.
14. Escoufier, Y.; Hayashi, C.; Fichet, B. *Data Science and Its Applications* [*La@science des données et ses applications*]. Academic Press/Harcourt Brace: Tokyo, 1995. ISBN: 0-12-241770-4. OCLC: 489990740.

15. Murtagh, F.; Devlin, K. The Development of Data Science: Implications for Education, Employment, Research, and the Data Revolution for Sustainable Development. *Big Data Cogn. Comput.* **2018,** *2* (2), 14. doi:10.3390/bdcc2020014.
16. Stewart T.; Tolle, K. M. *The Fourth Paradigm: Data-intensive Scientific Discovery*; Microsoft Research, 2009. ISBN: 978-0-9825442-0-4 (archived from the original on 20 March 2017; retrieved December 16, 2016).
17. CaoLongbing. Data Science. *ACM Comput. Surv.* **2017,** *50* (3), 1–42. doi:10.1145/3076253.
18. Wu, C. F. J. Statistics=Data Science?. Retrieved 2 April 2020.
19. Bell, G.; Hey, T.; Szalay, A. Computer Science: Beyond the Data Deluge. *Science* **2009,** *323* (5919), 1297–1298. doi:10.1126/science.1170411. ISSN: 0036-8075. PMID: 19265007.
20. Press, G. *A Very Short History of Data Science. Forbes.* Retrieved April 3, 2020.
21. Gupta, S. William S. Cleveland, December 11, 2015. Retrieved 2 April 2020.
22. *About Data Science | Data Science Association.* www.datascienceassn.org. Retrieved April 3, 2020.
23. Talley, J. ASA Expands Scope, Outreach to Foster Growth, Collaboration in Data Science. *Amstat News*, June 1, 2016. American Statistical Association.
24. Introduction: What Is Data Science? *Doing Data Science.* www.oreilly.com. Retrieved April 3, 2020.
25. Davenport, T. H.; Patil, D. J. Data Scientist: The Sexiest Job of the 21st Century. *Harv. Bus. Rev.* **2012**. ISSN: 0017-8012. Retrieved April 3, 2020.
26. Yau, N. Rise of the Data Scientist. *Flowing Data*; June 4, 2009. Retrieved April 3, 2020.
27. US NSF – NSB-05-40. *Long-Lived Digital Data Collections Enabling Research and Education in the 21st Century.* www.nsf.gov. Retrieved 3 April 2020.
28. Press, G. Data Science: What's The Half-Life of a Buzzword?. *Forbes.* Retrieved 3 April 2020.
29. American Statistical Association. ASA Statement on the Role of Statistics in Data Science. *Amstat News.* American Statistical Association. October 1, 2015 (archived from the original on 20 June 2019; retrieved 29 May 2019).
30. *Computer and Information Research Scientists: Occupational Outlook Handbook: U.S. Bureau of Labor Statistics.* www.bls.gov (retrieved April 3, 2020).
31. Nate Silver: What I Need from Statisticians. *Statistics Views.* www.statisticsviews.com (retrieved April 3, 2020).
32. *11 Data Science Careers Shaping the Future.* Northeastern University Graduate Programs, November 23, 2018 (retrieved April 3, 2020).
33. DharVasant. Data Science and Prediction. *Commun. ACM.* **2013,** *56* (12), 64–73. doi: 10.1145/2500499.
34. Pham, P. The Impacts of Big Data That You May Not Have Heard Of. *Forbes* (retrieved April 3, 2020).
35. Martin, S. How Data Science Will Impact Future of Businesses?. *Medium* **2019**. (Retrieved April 3, 2020).
36. Statistics Is the Least Important Part of Data Science. *Statistical Modeling, Causal Inference, and Social Science.* statmodeling.stat.columbia.edu (retrieved April 3, 2020).

37. Scott, S. M. *An Introduction to Python for Scientific Computing*, September 24, 2019 (retrieved April 2, 2020).
38. Granville, V. Data Science without Statistics Is Possible, Even Desirable. Blog, View, December 8, 2014 at 5:00 pm. www.datasciencecentral.com. Retrieved April 3, 2020.
39. Rhodes, M. A Dead-Simple Tool That Lets Anyone Create Interactive Maps. *Wired*, July 15, 2014. Retrieved 3 April 2020.

CHAPTER 3

Deep Learning: Current Trends and Techniques

BHARTI SHARMA[1*], ARUN BALODI[2], UTKU KOSE[3], and AKANSHA SINGH[4]

[1]*DIT University, Dehradun, India*

[2]*Atria Institute of Technology, Bangalore, India*

[3]*Suleyman Demirel University, Isparta/Turkey*

[4]*School of CSET, Bennett University, Greater Noida, India*

[*]*Corresponding author. E-mail: bharti.sharma@dituniversity.edu.in*

ABSTRACT

Deep learning is a subfield of artificial intelligence that is applied to solve the challenging and exciting problems in the various domains, for example, computer vision, 3-D object recognition, and natural language processing, etc. Diverse vendors (e.g., Google, Facebook, and Intel, etc.) have developed many frameworks to help the programmer to model the solution of the complex problem. Each of the deep learning neural networks includes distinguished feature. Broadly, the architectures of the deep learning are categorized into fully connected network, convolutional network and recurrent neural network. This chapter focuses on the architecture; frameworks of the most popular frameworks of deep learning with the objective to support end users make an informed choice about the best deep learning framework that suits their requirements and resources.

Deep Learning in Visual Computing and Signal Processing. Krishna Kant Singh, Vibhav Kumar Sachan, Akansha Singh & Sanjeevikumar Padmanaban (Eds.)

3.1 INTRODUCTION

Today, deep learning is measured as one of the warmest research disciplines in the domain of machine learning. Deep learning automatically works on supervised and/or unsupervised methods of the shallow structured learning architecture, rather than most traditional learning methods, to automatically learn hierarchical representations.[1] Deep learning designs can determine more complex input patterns than traditional learning methods. Two basic deep learning approaches have been proposed over the last decade. These methodologies are convolutional neural networks (CNN)[2] and deep belief networks (DBN),[3] which are well recognized in the domain of deep learning. Lately, deep learning has gained a lot of devotion in the investigation community as it performs well. It is also useful in many fields such as information fetching, natural language processing, and computer vision and so on. With the huge amount of data created around the world, which grows day by day, deep learning plays an important role in analyzing and forecasting results from this massive dataset. Expression is a central concept and idea in deep learning. Input abilities to outmoded machine learning algorithms need to be physically created from raw data to determine apparently interesting patterns, depending on the expertise and experience of the expert. The specialists require to fetch, analyze, select, and evaluate the features as well as creativity and some luck that make this process very time consuming. Deep learning techniques get their best function, especially from the information itself, but without human instruction. With the help of this the algorithm can discover the unknown or hidden features and relationship. The composition of simple data is used to show the complex data in the deep learning. The increasing complexity is handled by applying the idea of unsupervised hierarchical representations in the iterative deep learning. Deep learning is built on the artificial neural network (ANN) framework.[1,3] The purpose of the chapters in this book is to explain the deep learning framework and architecture as well as its evolution.

3.2 BACKGROUND

Machine learning offers the extensive set of procedures, including singular procedures and statistical approach such as Bayesian networks and linear, logistical regression approach. The singular and statistical algorithms like Bayesian, linear and logistical regression are the critical algorithms in machine learning. These algorithms have very good performance but in some situations like when learning go through with very complex and large datasets performance they get degraded. The neuron's cognitive, statistical, and learning algorithm along with a durable and permanent connection is the basis of the deep learning process.

The main focus of the computation process of neural network model is to generate a common neuron that can be useful for any type of statistics and can acquire knowledge in depth.[4] Deep learning is nowadays the most popular algorithm among researchers because deep learning makes learning procedure very easy because the deep learning algorithm has the ability to fetch the complex features automatically at high level of observations. This sophisticated learning feature of deep learning helps to solve the critical problems of computer vision. During the learning process in deep learning taking unlabeled data makes the training process exceptional.[5] Deep learning is a kind of multilayer architecture which includes multiple neurons in each layer of the network to solve the given problem of classification, regression, clustering, and so on. The each neuron which implements the activation function is known as logistic node. This logistic node is linked to the input of the succeeding layer and the weight at each node is adjusted using the loss function and according to the input data it is adjusted. Each layer of the neural network has many neurons which are allocated with different weights and contemporary try to discover the design in the input data.

In the multilayer architecture, each node learning from the output of the previous layer reduces the estimation to achieve the accurate output.[6] This process causes great complexity among some interconnected neurons. Initially, the word deep learning was applied by Igor and his team in the year 2000 in the context of artificial intelligence (ANN). Walter Pitts and Warren McCulloch were given the mathematical model of ANN which was published in the seminar entitled, "Logical Calculation of Managing Ideas of Nerve Function." Alexei Ivakhnenko and V.G. created the site including the in-depth learning network using theories and ideas in 1965. In between 1979 and 1980, Kunihiko Fukushima announced an artificial neural network that learned to identify the vision patterns. In the following years many authors also made the contribution. Hinton in 2006 asked how the brain works and emphasized on the concept of unsupervised training and deep beliefs networks after that in 2009, the ImageNet was announced by Fei-Fei Li and AlexNet was created by Alex Krishevsky in 2011. Geophint

Hinton and his team used the CNN in 2012. The authors observed that the error rate is dropped to 18.9%.[7] In 2014, Google proposed its own deep learning algorithms that are known as GoogLeNet. The performance of GoogLeNet is good as it reduced the error rate by 6.7%. After that deep learning is applied in the diverse fields like sentiment analysis,[8] machine translation and vocal language comprehension,[9] weather prediction, economic support, and much more.[10]

Nowadays, deep learning is a very powerful concept in the artificial intelligence domain. The impact was felt in all early scientific disciplines. It has already destroyed and transformed business and industry. There is competition to drive deep learning amongst the world's leading economic and technology companies. Deep learning is performing well beyond the human level in different domains. For example, predicting a movie rating, deciding whether to approve a loan application, or the time it takes to deliver a car.[11] The Turing Award that is Nobel Prize in computing is won by the three deep learners on March 27, 2019.[12] Currently, deep learning is doing well and a lot has been got but it still requires much more. Deep learning may improve the human life by accurate prediction especially in health domain, for example, accurate prediction of cancer disease,[13] finding of new medicines, and prediction of natural calamities.[14] For example, showed that the deep learning system was competent to diagnose 450 images of 2032 diseases at the equivalent level as a specialized dermatologist.[15] Google AI[16] was able to exceed the average accuracy of 70–61% in graded prostate cancer for general pathologists trained by the US Board of Directors. The some of the review papers on deep learning emphasized on detailed fields and operations, without including the full range of areas.[17–20]

3.3 ARCHITECTURE: NEURAL NETWORK AND DEEP LEARNING

Neural network can be categorized into feed-forward, recurrent, radial basis function, Kohonen Self Organizing and Modular Neural Network. The info flows in the forward direction from input through hidden nodes to the output layers in the feedforward neural networks. In this case there are no circles or loop backs formed. Contrasting neural networks of type feedforward, the RNNs computational units create the cycle. The output of the one layer is the feed to the subsequent layer. This is usually the single layer in the network; therefore, the result of the layer is a feedback to itself and it creates a feedback loop. This permits the network to hold retention about the earlier state and apply them to attract the existing output. One of the critical significances of this difference is that RNNs can get a set of inputs and produce a set of output values unlike feedforward. It is very important for the application that needs the computational sequence of the time series-based response data, special speech identification, and video classification frame by frame. For example, if an order of three-word sentences constitutes the input, each word corresponds to one layer and the network is expanded or expanded three times in a three-layer RNN.

The mathematical description is as follows at time t:x_t denotes the input. P, Q, and R are the learning parameters common in all steps and O(t) is the output. *St* denotes the state at time *t* and can be considered as follows. Where *f* is the stimulation function (such as ReLU).

$$S_t = f(Px_t + Rs_t - 1) \tag{3.1}$$

Generally, the classification, function fitting, and time series prediction problems is implemented using radial basis function. This type of neural network is denoted by an input layer, hidden layer, and an output layer. The hidden layer is implemented using the Gaussian function. Gaussian function is the kind of the radial basis function and each node represents the center of the cluster. The network trains to center its input and the output layer combines the output of the radial basis functions using the associated weight parameters to perform the classification.[24] The self-organization of models using input data is implemented using supervised learning in the Kohonen self-organizing neural networks. This network is made up of two fully connected input and output layers. The two dimension grid is used to implement the output layer and no activation function. The weights show the position of the node of output layer. The distance between input and output layer is calculated by the Euclidean distance to get the weights. The weights of the neighbors of the nearest node from the input data are modified using the following formula:[25]

$$ww_i(t+1) = ww_i(t) + \alpha(t)\mu_{j*i}(xx(t) - ww_i(t)) \tag{3.2}$$

Where, *xx* (t) denotes at time *t* the input data, ww_i (t) shows at time *t* the i-th weight and μ_{j*i} is the neighborhood function that works between node i^{th} and j^{th}. The modular neural network segments the large network into smaller independent neural network module. The small network does specific function after that the output of each small module is integrated as a single output of the entire network.[26] The Sparse Autoencoders,

Convolution Neural Networks (ConvNets), and Restricted Boltzmann Machines (RBMs) are the famous techniques used by the DNNs. Autoencoders are neural networks that perform dimension reduction by learning features or encodings from a particular dataset. The sparse auto encoder is a variation of the auto encoder where some units have values nearby to zero or inactive. Deep CNN uses several layers of unit groups that relate with inputs (or pixel values in the case of images) to extract the desired functionality. CNN has implemented in many fields like image processing recommender systems and in NLP. The learning of the probability distribution in the dataset is misused by the RB. The training is implemented by the back propagation in all the networks. The gradient descent is used to minimize the error by modifying the weights created on the partial derivative of the inaccuracy of each weight in the back propagation. The different categories are used to categories the neural network models.

1. Discriminative
2. Generative

In the discriminant model, **the data flows from the input layer through the hidden layer to the output layer.** It is a bottom-up approach. This model is used specially in classification, regression type of problems. Supervised learning uses this technique for training.

In the generative model data flows in the backward direction and it is a kind of top-down approach. The model is useful in the unsupervised pretraining and probability distribution problems. Given an input m and a corresponding label n, the discriminant model learns the probability distribution pd (n | m), that is, the probability of n where m is given directly.

A generative model, on the other hand, can learn the joint probabilities of pd (m, n) and predict P (n | m) from it.[27] In general, a criminal approach is taken to offer real training whenever labeled data are available and a generative approach when labeled data are not available.[28] The training of network can be done using supervised, unsupervised, and semi supervised approach. In the supervised learning, labeled datasets are used to train the network but in case of unsupervised approach unlabeled datasets are used to train the network so feedback is the best approach to train the network in this learning. In case neural networks are connected to the generative models like RBM, they can be fine-tuned using standard supervised learning. After that test dataset is used to find the pattern. It is then used in the test dataset to determine patterns or classifications. Big data has expanded the scope of deep learning with its enormous amount and diverse data. It is very hard to say which technique is better between supervised and unsupervised learning because both approaches have advantages and disadvantages and use cases. In some cases unsupervised learning showed better result in case of unstructured video sequences.[29]

Modified neural networks such as the DBN described by Chen and Lin[30] used both labeled and unlabeled data for both supervised and unsupervised learning to improve performance.

3.3.1 DNN ARCHITECTURES

3.3.1.1 THE DEVELOPMENT OF THE GPU AND DEEP LEARNING

Deep networks have different topologies take in deep learning. Deep learning includes several layers of networks for the practical application. Adding more layers shows more connections and weights between and within the layers. The GPUs is required to train and deploy the deep network. The GPUs which includes 1000–4000 specialized data processing cores works well in the case of deep network than conventional processor which includes 4–24 general purpose CPUs. The large number of processors makes the GPU more efficient by making the computations process more parallel than the conventional CPUs. This makes GPUs active for the huge neural networks that can compute several neurons at the same time. GPUs work well at floating point vector operation as neurons work based on vector multiplication and addition. Due to these features, the neural network works well on GPUs which is parallel in nature.

3.3.1.2 ARCHITECTURES OF DEEP LEARNING

There are different types of algorithms and architectures that are applied in deep learning. This section includes the five different types of deep learning architecture since the last 20 years. The most popular and frequently used algorithms in the deep learning are LSTM and CNN. Although both are the oldest algorithms, these designs are implemented in an extensive variety of situations and the following table includes some of their typical applications.

TABLE 3.1 Deep Learning Architectures and its Applications.[34]

Architecture	Use
RNN	Speech identification, handwriting identification
LSTM/GRU networks	Text compression using Natural language, recognition of handwriting, recognition speech, gesture and captioning of image
CNN	Recognition of image, processing of natural language, and analysis of video
DBN	Recognition of image, retrieval of information, understanding of natural language, prediction of failure
DSN	Retrieval of information, recognition of continuous speech

Recurrent Neural Networks

The deep learning algorithms are built on foundation of the RNN network. The main difference between a conventional multilayer network and recurrent network is that multilayer network is completely feedforward but in case of recurrent network there are connections that feed back into previous layers or into the same layer. To store the past input and to get the feedback, RNNs need memory. RNNs include the big set of instructions. For example, one popular technology, LSTM is an example. A key differentiator is feedback within the network. This becomes apparent from the hidden layer, the output layer, or some combination thereof. RNNs can be expanded in time and trained using typical back propagation or using a variant of back propagation called intrastate back propagation (BPTT).

LSTM/GRU Networks

LSTM has become an increasingly popular RNN architecture for various applications in recent years. It was proposed by Hochreiter and Schimdhuber in 1997. LSTM is included in everyday products such as smartphones. IBM has used LSTM to IBM Watson® for traditional voice recognition to set milestones. LSTM started from a classic neural network architecture and presented the idea of memory cells instead. A memory cell can hold its value as a function of input for short and long period. This permits the cell to remember what is important, not just the last calculated value. This permits the cell to hold the significant value rather than the last calculated value. LSTM memory cells used three gates to control the entering and exiting of the information in the cell. The new information flows into memory in control by the input gates. When the cell permits to hold new values ignoring the existing information then this function is controlled by forget gate. Ultimately, when the cell holds the information that is used in the output, in that case this function is controlled by the output gate. Each gate is controlled by the weights associated with the cell. The resultant network error is used to optimize these weights in the training algorithms. LSTM generalization was introduced in 2014. This model includes two gates and eliminates the output gates existing in the LSTM model. In many applications, GRU and LSTM have similar but simple means weight is less and execution is quicker. The GPU combines an update and a reset gate. The update gate shows the capacity that how much to hold the previous cell contents. The reset gate describes the inclusion of new value into prior cell contents. Standard RNN can be implemented by setting the reset gate to 1 and the update gate to 0 using GRU. GRU is simpler, faster to train, and more proficient in its execution than LSTM. Though LSTM is more communicative and the more data you have, the better results you will get.

Convolutional Neural Networks (CNN)

CNN is biologically stimulated by the visual cortex of animals. It comes under the category of multilayer neural network. This neural network design is suitable in the image-based computer vision kind of applications. The leading CNN architecture was created by Yann Le Cun. The CNN at that time worked on the script identification and applied in the postal code interpretation. Initially, the deep network can only identify the basic features like edge but after that fetched features project into advanced level attributes of the input. The feature extraction and classification is implemented using many layers in the LeNet CNN. The image is segmented into receptive fields and then sent to the convolution layer to extract features from the input image. The max pooling is used to retain the important information and down sampling is applied to minimize the size of the extracted features then again convolution and pooling is done to feed into fully connected multilayered perceptron. The extracted features of an image are represented by the output layer of this network. The back propagation is used to train the network. The use of fully connected input, hidden and output layer with convolution and pooling concept in the deep layer has help investigators to develop the new applications for deep learning neural network. The CNN is not only working well in the image detection but also developing good applications in natural language processing. The image and video capturing systems where images or videos are summarized in natural language use recent application of CNN and LSTM. Video processing is implemented using CNN and the CNN output is translated into natural language in case LSTM is used for training.

Deep Belief Networks

The deep belief is a distinctive system design including new training algorithms. A DBN is a deep network including multiple hidden layers and each twosome of linked layers is controlled by RBM. Stack of RMBs is used to represent the DBN. In the DBN, the raw sensory input is used to denote the input layer and hidden layer gets an abstract denotation of this input. The output layer which is controlled slightly in a different way than the other layers deployed network classification. Supervised and unsupervised techniques are the techniques used for the training.

In the unsupervised pretraining, the RBM is taught to restructure its input. Initially, the input layer is reconstructed into the first hidden layer in the RBM. The same process is implemented to teach the remaining RBM but in this case the first hidden layer is treated as the input layer and RBM is taught using the output of the first hidden layer. This method is continued until each layer is pretrained. After the completion of the pretraining, the fine tuning will be started. In this phase, label is applied to the output nodes and assigned them meaning in the context of the network. The training process is completed using either gradient descent learning or back propagation.

Deep Stacking Networks

The deep convex network is a final network that is represented as DSN. There is a difference between DSN and traditional deep learning network. Although a DSN is represented as a combination of deep networks, it's a deep set of distinct nets in reality each with having its own hidden layer. The deep learning complexity is addressed by the structural design. In the deep learning design, each layer training complexity increase exponentially, so the DSN takes teaching as a series of distinct training problem rather than a single problem.

A DSN is made up of sequence of modules and each module represents a subpart of the network in the hierarchy of the DSN. The architecture of DSN has been created using three components. Every component of architecture combines an input, an output, and one single hidden layer. The components are piled using the stack pattern and each component takes the input from the prior layer and the original input vector. The multilayer architecture permits to train the complex classification across the network than if you use single module. With DSNs, you can train individual modules individually, which makes it efficient when parallel training is possible. The back propagation technique is used as supervised learning in each component instead of using back propagation in the whole nets. In many applications, DSNs outperform common DBNs, making them a widespread and competent network architecture.

3.4 DEEP LEARNING ARCHITECTURES

While it's certainly likely to deploy these deep learning architectures, it is very time taking and requires time to optimize and mature. Fortunately, several open source frameworks are available to make implementing and deploying deep learning algorithms easier. These frameworks support languages such as Python, C/C++, and Java® languages. Let's take a look at the three most popular frameworks and their pros and cons.

3.4.1 CAFFE

Caffe is the frequently used deep learning framework. It was developed during the PhD of an academician. A paper, currently released under the Berkeley Software Distribution license.

One of the frequently used deep learning frameworks is known as Caffé. The CNN and LSTM are supported by Caffe and RBM and DBM are not supported by Caffe but the future release of Caffe, that is, Caffe2 supports RBM and DBM. NVIDIA CUDA deep neural network library is applied to support the classification of the image, and to support the acceleration based on GPU. The deep learning algorithms are parallelized on a cluster of systems by open multi-processing which is supported by Caffe. C++ is used for the implementation of the Caffe and Caffe2 and for the training and execution of deep learning, Python and Matlab are used as an interface.

3.4.2 DEEPLEARNING4J

Deeplearning4j is implemented using java technology and it is a very famous and useful deep learning framework and for working on other languages like Scala, Python, it uses application programming interface. Apache license is used to release the framework and it also supports RBM, DBN, CNN, and RNN. Apache Hadoop is used to support the distributed parallel version and Spark is used for the Big Data Processing Framework in the deeplearning4j. Deeplearning4j has been used to solve different problems like fraud detection, recommender system, image recognition, cyber security (network intrusion detection) in the financial sector. The GPU optimization is done using CUDA and Open Multi-Processing is used to include the parallelism in the framework.

3.4.3 TENSORFLOW

TensorFlow is developed by Google as a descendant of the closed source DistBelief. It is released by the Apache 2.0 and license of it is freely available. The different types of networks like CNN, RBM, DBN, and RNN can be trained and deployed using TensorFlow. Image captioning, malware detection, voice recognition, information retrieval problems can be implemented using TensorFlow. Recently, an android-specific stack is released which is named as TensorFlow Lite. Python, C++, Java language, Rust, or Go can use TensorFlow to develop the application. Python is more stable and distributes its execution in Hadoop. In addition to special hardware interfaces, TensorFlow supports CUDA

3.4.4 DISTRIBUTED DEEP LEARNING

The distributed deep learning (DDL) is known as jet engine for deep learning and the major framework such as Caffe and TensorFlow are linked by IBM DDL. The deep learning algorithms can increase their performance on the hundreds of GPUs and cluster of servers using DDL algorithm. The optimal path is calculated between GPUs is assigned to resulting data and uses to optimize the communication of neuronal computations in the DDL. The bottleneck of the solution of the deep learning cluster was demonstrated by defeating an earlier image recognition task recently set by Microsoft.

It is difficult to develop a method to automatically extract meaningful features from labeled and unlabeled high-dimensional data spaces. So, supervised and unsupervised approach has integrated to achieve this.[31] The semi supervised learning techniques are defined as the complement of the unsupervised and supervised techniques .The early convergence and the over fitting are the two main issues in the DNNs that need to resolve.

Early convergence happens when the weights and biases of the DNN settle at an optimum only at the local level, overlooking the global minimum over the multidimensional space. Over fitting, on the other hand, is robust and less adaptable because the DNN is fine-tuned to a particular training dataset and does not fit into other test datasets. There are many machine learning frameworks and libraries that make the training model easier apart from training algorithms and architectures. With these frameworks, you can leverage difficult mathematical functions, training procedures, and statistical modeling without having to write your own. Some have distributed and parallel processing capabilities and suitable improvement and deployment proficiencies. GitHub is the largest source code hosting service provider in the world.[32] GitHub star indicates the popularity of the project on GitHub. TensorFlow is the most popular DL library.

KEYWORDS

- **machine learning**
- **neural networks**
- **deep learning**
- **architecture**

REFERENCES

1. Qiu, J.; Wu, Q.; Ding, Q.; Xu, Y.; Feng, S. A Survey of Machine Learning for Big Data Processing. *EURASIP J. Adv. Signal Process.* **2016,** *1*(7).
2. Le Callet, P.; Viard-Gaudin, C.; Barba, D. A Convolutional Neural Network Approach for Objective Video Quality Assessment. *IEEE Trans. Neural Netw.* **2006,** *17*(5), 1316–1327.

3. Yu, D.; Deng, L. Deep Learning and its Applications to Signal and Information Processing [Exploratory dsp]. *IEEE Signal Process. Mag.* **2011,** *28*(1), 145–154.
4. Ganatra, N.; Patel, A. Comprehensive Study of Deep Learning Architectures, Applications and Tools. *Int. J. Comput. Sci. Eng.* **2016,** *6*(12).
5. Nielsen, M. Neural Networks and Deep Learning, 2017. http: //neuralnetworksand-deeplearning.com/.
6. Surajit Chaudhuri, U. V. An Overview of Business Intelligence Technology. *Commun. ACM* **2011,** *54*(8), 88–98.
7. Patel Sanskruti, A. P. Deep Leaning Architectures and its Applications: A Survey. *Int. J. Comput. Sci. Eng.* **2018,** *6*(6), 1177–1183.
8. Krizhevsky, A.; Sutskever, I.; Hinton, G. E. ImageNet Classification with Deep Convolutional Neural Networks. *Adv. Neural Inform. Process. Syst.* **2012,** 1097–1105.
9. Socher, R.; Perelygin, A.; Wu, J. Y.; Chuang, J.; Manning, C. D.; Ng, A. Y.; Potts, C. Recursive Deep Models for Semantic Composition- Ality Over a Sentiment Tree-bank. *In* Proceedings of the Conference on Empirical Methods in Natural Language Processing, Citeseer, 2013.
10. Mesnil, G; Dauphin, Y.; Yao, K.; Bengio, Y.; Deng, L.; Hakkani-Tur, D.; He, X.; Heck, L.; Tur, G.; Yu, D.; Zweig, G. Using Recurrent Neural Networks for Slot Filling in Spoken Language Understanding. *IEEE/ACM Transac. Audio Speech Lang. Process.* **2015,** *23*(3), 530–539.
11. Le, Q. V. In *Building High-Level Features Using Large Scale Unsupervised Learning.* IEEE International Conference on Acoustics, Speech and Signal Processing, 2013.
12. Ng, A. Machine Learning Yearning: Technical Strategy for AI Engineers in the Era of Deep Learning. *Tech. Rep.* **2019.**
13. Metz, C. *Turing Award Won by 3 Pioneers in Artificial Intelligence*; New York Times: New York, NY, USA, 2019; p B3.
14. Nagpaletal, K. Development and Validation of a Deep Learning Algorithm for Improving Gleason Scoring of Prostate Cancer. *CoRR.* **2018.**
15. Nevo, S. ML for Flood Forecasting at Scale. *CoRR.* **2019.**
16. Esteva, A.; et al. Dermatologist-Level Classification of Skin Cancer with Deep Neural Networks. *Nature* **2017,** *542*, (7639), 115.
17. Arulkumaran, K.; Deisenroth, M. P.; Brundage, M.; Bharath, A. A. Deep Reinforcement Learning: A Brief Survey. *IEEE Signal Process. Mag.* **2017,** *34*(6), 26–38.
18. Gheisari, M.; Wang, G.; Bhuiyan, M. Z. A Survey on Deep Learning in Big Data. *Proc. IEEE Int. Conf. Comput. Sci. Eng. (CSE).* **2017,** 173–180.
19. Pouyanfar, S. A Survey on Deep Learning: Algorithms, Techniques, and Applications. *ACM Comput. Surv.* **2018,** *51*(5), 92.
20. Vargas, R.; Mosavi, A.; Ruiz, R. Deep Learning: A Review. *In Proc. Adv. Intell. Syst. Comput.* **2017,** 1–11.
21. Hoffmeister, L. V.; Moura, G. M. S. S. d. Use of Identification Wristbands Among Patients Receiving Inpatient. *Revista Latino-Americana de Enfermagem (RLAE)* **2015,** *23*(1), 36–43.
22. Jones, M. T. index.html IBM, 08 September 2017. [Online]. https://www.ibm.com/developerworks/library/cc-machine-learning-deeplearning-architectures/index.html (accessed Jan 17, 2018).

23. Rani, K. U. Analysis of Heart Diseases using Neural network Approach. *Int. J. Data Mining Knowl. Manag. Process* (IJDKP) **2011,** *1*(5), 1–8.
24. Buhmann, M. D. *Radial Basis Functions*; Cambridge Univ. Press: Cambridge, U. K., 2003; p 270.
25. Akinduko, A. A.; Mirkes, E. M.; Gorban, A. N. SOM: Stochastic Initialization Versus Principal Components. *Inf. Sci.* **2016,** *364–365*, 213–221.
26. Chen, K. Deep and Modular Neural Networks. *In Springer Handbook of Computational Intelligence*; Kacprzyk, J., Pedrycz, W., Eds.; Springer: Berlin, Germany, 2015; pp 473–494.
27. Ng, A. Y.; Jordan, M. I. In *On Discriminative vs. Generative Classifiers: A Comparison of Logistic Regression and Naive BAYES. Proceedings of 14th International Conference on Neural Information Processing Systems, Cambridge*, MA, USA: MIT Press, 2001; pp 841–848.
28. Bishop, C. M.; Lasserre, J. Generative or Discriminative? *Getting the Best of Both Worlds. Bayesian Stat.* **2007,** *8*, 3–24.
29. Zhou, T.; Brown, M.; Snavely, N.; Lowe, D. G. Unsupervised Learning of Depth and Ego-Motion from Video. *CoRR*. **2017**.
30. Chen, X. W.; Lin, X. Big Data Deep Learning: Challenges and Perspectives. *IEEE Access* **2014,** *2*, 514–525.
31. LeCun, Y.; Kavukcuoglu, K.; Farabet, C. Convolutional Networks and Applications in Vision. *Proc. IEEE Int. Symp. Circuits Syst.* **2010,** 253–256.
32. Gousios, G. B.; Vasilescu, A. S.; Zaidman, A. In *Lean GHTorrent: GitHub Data on Demand*, International Proceedings 11th Working Conference on Mining Software Repositories, Hyderabad, India, 2014; pp 384–387.
33. LeCun, Y.; Bengio, Y.; Hinton, G. Deep Learning. *Nature* **2015,** *521*(7553), 436–444. https://developer.ibm.com/articles/cc-machine-learning-deep-learning-architectures/
34. Arora, M. S.; Singh, D. K. Deep Learning: Overview, Architecture, Framework & Applications. *Int. J. Latest Trends Eng. Technol.* *10*(1), 379–384.
35. Trends, Google, [Online]. https://trends.google.com (accessed Jan 7, 2018).

CHAPTER 4

TensorFlow: Machine Learning Using Heterogeneous Edge on Distributed Systems

R. GANESH BABU[1*], A. NEDUMARAN[2], G. MANIKANDAN[3], and R. SELVAMEENA[4]

[1]*Department of Electronics and Communication Engineering, SRM TRP Engineering College, Trichy, TN, India*

[2]*Department of Electrical and Computer Engineering, Kombolcha Institute of Technology-Wollo University, Ethiopia*

[3]*Department of Electronics and Communication Engineering, Dr. M.G.R. Educational and Research Institute, Chennai, TN, India*

[4]*Department of Computer Science and Engineering, Dr. M.G.R. Educational and Research Institute, Chennai, TN, India*

[*]*Corresponding author. E-mail: ganeshbabu.1509@gmail.com*

ABSTRACT

TensorFlow is an edge for imparting AI estimations, and a utilization for the stage counts this way. An estimation imparted utilizing TensorFlow should be possible with for all intents and purposes zero changes in a wide scope of heterogeneous systems, going from Personal Digital Assistants, for instance, phones and tablets, to monstrous scope proper structures of numerous PCs and countless computational devices, for instance, GPU. The system is versatile and be able to be utilized to impart a large scope

Deep Learning in Visual Computing and Signal Processing. Krishna Kant Singh, Vibhav Kumar Sachan, Akansha Singh & Sanjeevikumar Padmanaban (Eds.)

of figuring, counting the planning and inference of computations for profound neural system representation, and has been utilized to manage the investigation with sending of AI structures to in excess of 12 programming designing zones and various fields including talk acknowledgment. It is sight of PCs, electronic innovation, information recuperation, regular language dealing with and recovery of spatial information, and revelation of gadget prescription.

4.1 INTRODUCTION

In Google Brain, movement began to explore the usage of colossal degree of significant neural systems for equally divulgence and use in the Google things.[1] As a bit of the early work at the present time, DistBelief, our interesting versatile orchestrating and acknowledgment organization, has been gathered and this system served us glowing. Designers and those at Google used DistBelief to test research for solo language depiction learning, photo solicitation and article affirmation templates, video gathering, movement directing talk affirmation, moving location, fortification learning, and distinctive regions.[16] However, in close interest to the Google Brain genius, more than 50 Google gatherings and numerous letters, all together, have routinely passed on important neural systems using DistBelief in a wide range of ways, including Google Search of our company, our conversation affirmation frameworks, Google Photos, Google Maps and Google Images, Google Translate, Twitter, and various other things.

Tensor stream figurings are passed on as state-of-the-workmanship data stream charts (envisioned in more detail in Section 4.2), and we have focused on making the structure both adaptable for brisk examination of new models for assessment purposes and satisfactorily unmatched and vigorous for the course of action and sending of AI models. To scale up the neural framework preparing for greater strategies, TensorFlow engages customers to capably confer various kinds of parallelism through the replication and sensible execution of an inside model dataflow layout, with a wide arrangement of computational gadgets all collaborating to animate a couple of ordinary boundaries or various conditions. Unassuming changes in the figuring depiction grant a wide grouping of ways to deal with management parallelism to be polished and tried with low exertion. Some TensorFlow uses license of some adaptability to the degree boundary revives are consistent, and we can beyond question pass on and misuse these nice synchronization basics in a segment of our greater affiliations.

Stood out from DistBelief, TensorFlow's changing model is continuously versatile, its presentation is on a very basic level better, and it supports getting ready and using a more broad extent of models on a more broad arrangement of heterogeneous hardware stages. Our numerous inward clients of DistBelief have quite recently changed to TensorFlow. These clients rely upon TensorFlow for innovative work, with endeavors as varying as running conclusions for PC seeing models on mobile phones to get ready immense degree for profound neural frameworks with a few billion boundaries on a huge number of model records utilizing countless machines. In spite of the way that these applications concentrated explicitly on AI and profound neural frameworks, we expect that the impressions of TensorFlow resolve to be valuable in an assorted variety of various spaces, tallying different kinds of AI figurings and potentially various sorts of numerical counts.

The rest of this chapter portrays TensorFlow in more prominent detail. Section 4.2 portrays the TensorFlow interface programming model and significant ideas, and Section 4.3 delineates our single machine just as our coursed executions. Section 4.4 diagrams a few extensions of the fundamental programming reproduction and Section 4.5 frameworks a few advances to the basic executions. Section 4.6 shows a part of our encounters when utilizing TensorFlow, Sections 4.7 and 4.8 delineate a couple of programming interesting expressions that consider accommodating when utilizing TensorFlow, and Section 4.9 outlines a couple of right-hand apparatuses that worked around the center framework for TensorFlow. Independently, Section 4.10 discusses potential and related work and Section 4.11 offers the final details.

4.2 BASIC PROGRAMMING PERCEPTION AND REPRESENTATION

TensorFlow estimation is spoken through an organized outline, which comprises a great deal of center points. The chart addresses an information stream gauge, with additions to permit a couple of sorts of center points to keep up and update mechanical conditions and to broaden and circle compose structures inside the graph in a technique like Naiad. Clients make a consistent graph all the time utilizing single of the keep-up frontend lingos. A method part for the development and execution of a TensorFlow graph using the python face-end is revealed in Figure 4.1 and the count chart is in Figure 4.1. The core has at any rate zero sources of information and zero yields in a TensorFlow graph and addresses the start of an action. Attributes that stream along the chart's mill edges (inputs from respects) are tensors, self-decisive bunches of dimensionality that assess or instigate the critical segment structure at the time of diagram formation. The outline also contains uncommon edges, known as control conditions: no information streams alongside those edges, but it states that the center point of control dependency establishment must be implemented before the goal center point of control dependency starts to execute. Because our model integrates variable status, customers may use the control conditions legally to allow the opportunity before associations occur. Its execution regularly coordinates organized rules for requesting self-sufficient activities regardless as an instrument for checking, for instance, the utilization of apex memory.

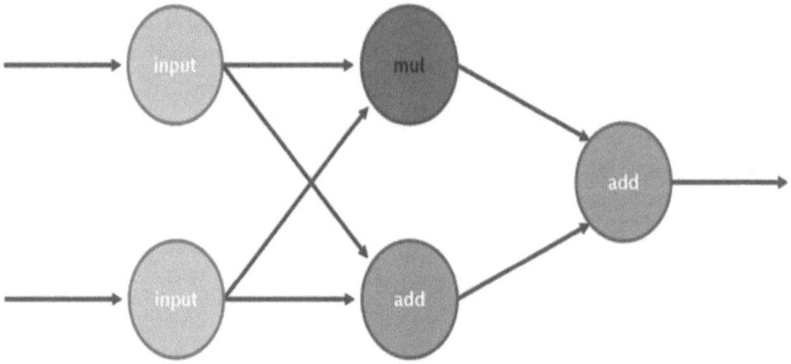

FIGURE 4.1 Consequent statistical illustration.

4.2.1 BEHAVIOR AND KERNELS

Every activity has a name and a scientific estimation (e.g., "matrix copy" or "incorporate") is indicated. A movement may have qualities, and all attributes must be given or derived at the hour of making of the diagram so as to begin an inside for the action to occur. One straightforward utilization of attributes is polymorphism assignments across various sorts of tensor parts. A piece is a demanding activity execution that is ready to be played out at a particular classification of contraption.

4.2.2 TENSORFLOW SYSTEM

Through holding a session with the TensorFlow program, company administrations communicate. The session interface underpins an extended approach for broadening the current outline administered by meeting additional center points and edges to create an estimate map (the basic graph is unfilled when a gathering is made). The supplementary significant activity kept up by gathering in which get a ton of yield to prepare, much the same as an optional tensor course of action to be dealt with in the guide rather than exceptional centers yields. Utilizing the motivations to utilization for the TensorFlow be able to speak the transitive terminate of the considerable number of centers that must be executed so as to record the yields that have been expressed and would then be able to arrange the execution of the right centers in a solicitation that regards their conditions.

4.3 IMPLEMENTATION

The key parts of a TensorFlow association are the customer who uses the meeting limit to resolve the ace within any case one form of worker, with each specialist technique responsible for interfering in any case with one gadget contract and for updating the graph center points on the devices of the person. If the two have captured neighborhood and use of the TensorFlow interface, near to use is used when the client, the ace, and the expert from nowhere are common for a single PC with regard to a specific working system stage (possibly with clear devices if, for example, separate GPU cards have been used in the machine). The scattered utilization confers the main piece of code to the zone's execution but broadens it with the guide of a circumstance wherein the customer, the expert as well as the staff could all run different machines in different systems.

4.3.1 TENSORS

A tensor is an introduction to our execution generated multi-dimensionally. We reinforce a variety of tensor modules like checked and unsigned whole numbers running in size from 8 pieces to 64 pieces, IEEE coast and double types, an unprecedented number sort, and a string type (a possible byte show-up). Support store of the correct size is truly handled by allocator to the tensor harps on the gadget. For comparison purposes, tensor sponsorship shops are tracked if no references remain.

4.3.2 EXECUTION OF SINGLE DEVICE

Next, it causes one to consider the least complex implementation circumstance: a solitary proactivity with a lone contraption. The chart centers are actualized in an application that identifies with the conditions between center points. As a matter of fact, we track each center point a mean of the quantity of conditions not yet implemented at that center point. The center point will be ready for execution and will be attached to a ready line when this check drops to zero. In some obscure request, the ready line which assigns the execution of the piece for a center point to the object of the contraction is dealt with. The ensure of all center points that rely upon the completed center point be diminished exactly when a center is wrapped up.

4.4 EXECUTION OF MULTIDEVICE

There are two basic complexities when a framework has different devices: picking the device to situate the estimation for every center point in the graph and afterward managing the vital correspondence of information over the restrictions of the contraption inferred by these decisions of circumstance.

4.4.1 NODE ASSIGNMENT

One of the basic duties of TensorFlow implementation is giving a computation map to portray figuring on the course of action of available contraptions. With respect to bolstered by this gauge, one contribution to locating estimation is a charge representation that includes evaluations of data spans and brings tensors with each graph center point, as well as evaluations of the figure time needed for each center given their data tensors. The representation of this charge is either strongly evaluated as endorsing heuristics related to specific behavior or is determined reliant on a certified course of action of position decisions for earlier chart implementation.

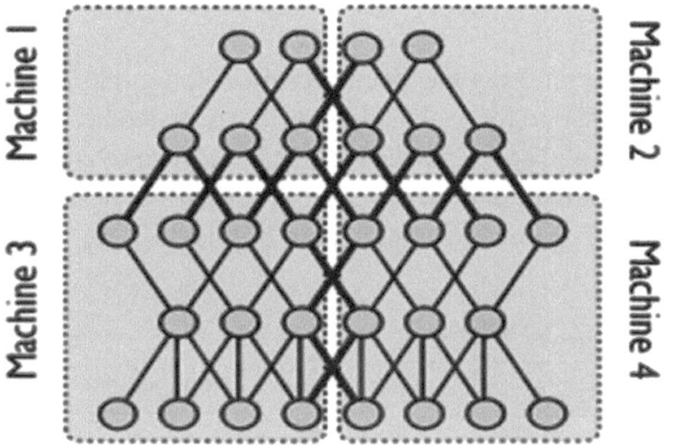

FIGURE 4.2 Single machine and distributed system structure.

The multiplication is demonstrated as follows and ends up choosing a device for every center in the guide utilizing insatiable heuristics. The center point to device structure made by this reenactment is additionally utilized as the genuine execution circumstance. The area figuring starts with the estimation graph's wellsprings and reproduces the procedure on every device as it advances. For every center that has quite recently shown up, a determination of achievable devices is thought of (a contraption may not be attainable if the device doesn't give the particular movement a piece). Among centers with explicit feasible contraptions, the circumstance figuring utilizes an anxious heuristic that takes a gander at the ramifications of putting the center point on any possible device for the time of fulfillment. This heuristic gets into clarification the deliberate or anticipated usage time of method on this type of contraption from the charge portrayal and furthermore includes the consumption of a few correspondences it will all be acquainted with communicating contributions from different devices to the device thought. The device where the activity of the center is to be done at the earliest opportunity is chosen as the device for that movement and the circumstance technique at that stage starts to choose the course of action decisions for the different centers in the graph, including downstream center points right now being set up for their own resanctioning. A couple of increments permit customers to offer bits of knowledge and fragmentary objectives to deal with the computation of the circumstance. The estimation of the understanding is a zone of advancement inside the framework.

4.4.2 CROSS-DEVICE COMMUNICATION

Once the focal point area has been recruited, the guide gets divided into several subcharts, one for each gadget. A few cross-contraction edges from x to y are discharged and replaced by an edge from x to another send focus point in the x segment and an edge in the suboutline from a receive focus highlight y. See Figure 4.3 for an update of this drawing.

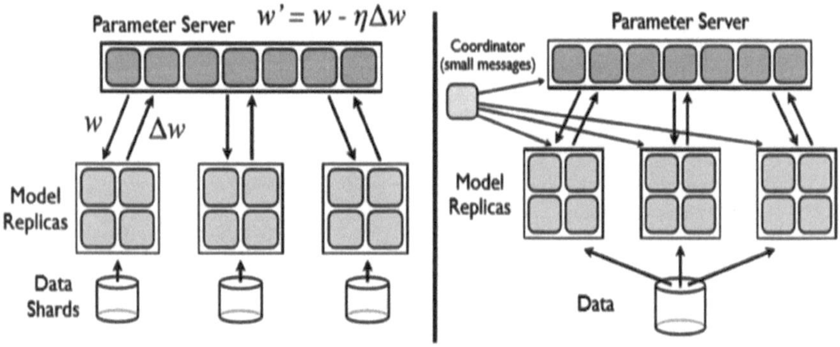

FIGURE 4.3 Message send/receive nodes are inserted asynchronously get parameters.

Throw and receive centers execution sketch for moving data through devices at dash time.[2] This encourages us to expel all correspondences inside send and get executions, hence improving the reprieve of the runtime.[6] When include throw and its receive centers consecrate all clients of a demanding tensor, a particular device is used at a specific receive center point rather than one receive center point on a particular device for each downstream client. This guarantees the data are just throwing once among a source contraption for the fitting tensor! Objective contraption pair and the tensor recollection is done only once for the objective device. By adapting to correspondence at this moment, it is additionally conceivable to decentralize booking of the individual center points of guide on explicit devices into the laborers: the center points throwing and receive award imperative correspondence among various authorities with contraptions just needs to give a single one.[7] Dash interest for each chart execution to each expert who has any center points for the graph, instead of taking an interest in the readiness of every center point or correspondence with each cross-device. It makes the edge extensively more versatile and takes into account much better tasks of the granularity place than if the arrangement must be finished.

4.4.3 DISTRIBUTED EXECUTION

Suitable execution of a chart basically is equivalent to execution of multi-devices. A sub-graph is delivered per contraption after area of the gadget.[9] Send/Get center matches that utilization distant correspondence instrument across authority types, for example, TCP or RDMA, to move data across framework limits.

4.5 OPTIMIZATIONS

Consider a segment of the TensorFlow adjustments that improve the execution or use of system properties.

4.5.1 GENERAL ELIMINATION OF SUBEXPRESSION

While computation graphs are every now and again made by an extensive scope of suggestion layers in the client code, figuring outlines will without a doubt finish up with abundance copies of a comparable estimation.[8] To fix this, we have changed a standard subexpression move, for example, the count which runs bigger than the computation chart and consecrates different copies of assignments with undefined wellsprings of data and class of activity to just one of these center points, and the divert edges appropriately speak to this canonization.[15]

4.5.2 DATA COMMUNIQUÉ AND OVERPROTECTIVE OF THE MEMORY USAGE

Mindful arrangement of TensorFlow exercises will prompt enhanced execution of framework, explicitly with respect to affecting data and utilizing memory.[11] Specifically, reserving time through the transitional results ought to be held in a memory in exercises and subsequently the utilization of apex memory. This lessening is especially noteworthy for contraptions with

GPU where the memory is constrained. Likewise, sorting out information correspondence through gadgets will bring down arguments about the game plan of properties. This includes receive center points intending to scrutinize distant resources. In the event that no protection is taken, these center points will begin a lot of sooner than would as a rule be worthy, likely simultaneously as execution starts. By playing a consider quick-as-time grants/as-late-as could be expected under the circumstances (ASAP/ALAP) as regular in exercises are broke down, we separate the fundamental outline approaches to decide when to make the get center points.

4.5.3 ASYNCHRONOUS KERNELS

Our structure also wires nonblocking bits, except traditional synchronous pieces which complete implementation toward the completion strategy. Its portions use slightly extraordinary boundaries to transfer a continuation of the calculation technique that should be conjured up once the bit is executed. It is an improvement in situations where multiple dynamic strings are usually costly when it comes to memory use or different properties and lets us avoid tying up an endless timeline execution string while sitting tight for I/O or various occasions. Noncompetitive instances combine the receiving portion and the enqueue and dequeue parts.

4.6 EXPERIENCE AND CONDITION

Under an Apache 2.0 grant, the TensorFlow execution and a sample implementation were freely distributed, and the software can be downloaded from www.tensorflow.org. The stage gives lumpy documentation, various preparatory exercises, and different models agent how to use the association for a wide range of AI tasks. The simulations coordinate layouts for collecting digits from the MNIST dataset (the "welcome universe" of AI calculations) arranging images from the CIFAR-10 dataset, showing occasional language association LSTM, getting ready word embedding vectors and that's just the starting stage.

The execution of the TensorFlow and an example use were openly delivered under an Apache 2.0 license, and the product can be downloaded at www.tensorflow.org. The stage gives dirty documentation, various preparation exercises, and different models to delegate how to use the association for a wide array of AI tasks. The designs coordinate templates for collecting digits from the MNIST dataset (the "welcome universe" of AI calculations) (Ganesh Babu et al., 2020), arranging images from the CIFAR-10 dataset, showing intermittent language LSTM[12] interaction, having ready word embedding vectors, and that's just the starting point.

4.7 MODEL

4.7.1 MODEL EQUIVALENT TRAINING

Model-equivalent getting-ready different pieces of the model estimation are done on different computational contraptions simultaneously for a comparative cluster of models that is in like manner easy to impart in TensorFlow.[17] An instance of irregular, significant LSTM representation used for plan to gathering learning and parallelized crossways three one-of-a-kind devices. Concurrent steps for model computation pipelining another customary technique to give indications of progress use for getting ready significant neural frameworks are pipeline the figuring of the representation inside comparable devices, by the running scarcely any synchronous steps inside a comparative plan of devices.[19]

This is showed up in Figure 4.8. It is genuinely similar to no simultaneous data parallelism; on the other hand, really the parallelism occurs inside the identical device(s), rather than reproducing the estimation outline on different devices.[3]

4.8 APPEARANCE

For this chapter, the future variation is to finish the introduction evaluation territory of both the single machine and the dispersed utilization.[10] Resources in this section describe a few instruments that we have built that are located near the middle TensorFlow object model.

4.8.1 TENSOR BOARD

Perception of graphic structures and overview of knowledge have produced TensorBoard, an accomplice portrayal gadget for TensorFlow, which is associated with the open-source release, to help customers understand the structure of their estimation outlines and the overall direct of AI models.[14] Portrayal of computation graphs of some of the figuring diagrams can be highly unpredictable for important neural frameworks. For example, the estimation graph for setting up a model such as the Google inception model, a significant convolution for best request execution in the Image Net 2014 test, has a computation outline of more than 37,000 center points TensorFlow and some significant irregular LSTM representations for showing contain more than 16,000 center points.

In light of the geography graphs, artless portrayal techniques much of the time produce confused and overwhelming frameworks. To help customers see the core relationship of the graphs, the Tensor Board dissolution center computations point into elevated scale squares, highlighting packages of ambiguous structures.[18] The structure moreover detaches out serious extent center points, which every now and again serve bookkeeping limits, into an alternate domain of the screen. Doing so diminishes visual wreck and focuses thought on the middle portions of the figuring diagram. The entire recognition is instinctive: customers can skillet, zoom, and stretch out assembled center points to exhaust down for nuances. An instance of the discernment for the outline of a significant convolution picture portrayal is showed up in Figure 4.4. Impression of the rundown of knowledge being ready for AI models, customers regularly have to select after a while to analyze the status of different parts of the representation to adjustments, supporting a series of specific description assignments on the inserted outline to end the TensorFlow.

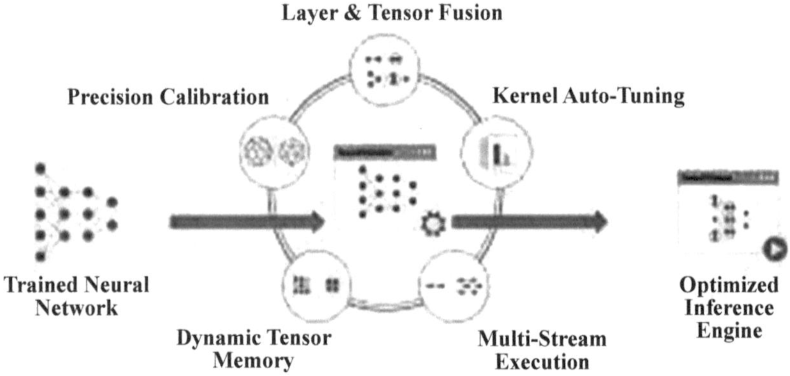

FIGURE 4.4 Tensor board graph representation of a convolution neural system model.

Regularly count graphs are set up so that summary center points are used to screen distinctive interesting characteristics, and now and again during execution of the readiness diagram. The course of action of summation center points is moreover executed, despite the run-of-the-mill plan of center points that are implementing. The client operator program creates the layout data to a log report that is related to model planning when the portrayal gets ready. The tensor board system is then worked on to display this log report for new overview data and it can demonstrate the details in the model and how it changes after a few. A screenshot of tensor board's one-off depiction shows up in Figure 4.5.

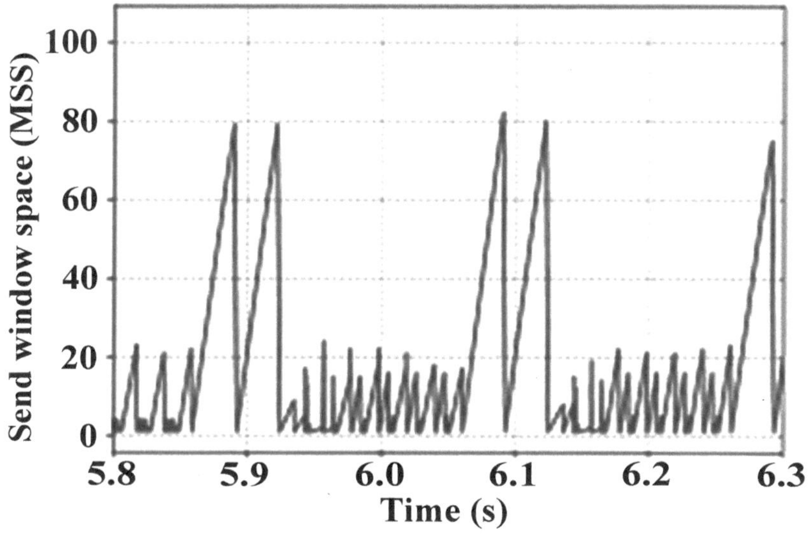

FIGURE 4.5 Tensor board visual overview of model description time-series data statistics.

4.8.2 PRESENTATION ANALYSIS

Any major DMA-related deferrals due to touch or eases back down are isolated and seen using jolts in observation. The UI gives a schematic of the entire system from the outset, showing only the most outstanding remnants of the introduction. While the consumer is continually zooming in, distinctions are slowly being rendered to fine targets. A PC EEG experience of a model is being set up on a multifocal CPU stage. The top third of the screengrab displays TensorFlow errands that are executed in equal amounts, as the dataflow impediments appear. The following base area demonstrates how most activities are turned into specific working environments when simultaneously acting in a string pool. The corner-to-corner jolts demonstrate where lining pause occurs at right-hand stage in the string bowl. Figure 4.6 demonstrates another EEG representation, with the GPU calculation mostly occurring. The host strings seen cooperating with TensorFlow GPU commands as they become executable and the housekeeping strings concentrate around the processor in various shades. In fact, jolts show where strings are eased back to CPU developments on GPU, or where activities experience gigantic lines-by-line postponement. Finally, Figure 4.7 provides a nitty-gritty understanding that assists in checking how different GPU streams are allocated by TensorFlow GPU over servers. Each time we try to unveil the difficult prerequisites in the information stream graph of GPU contraption, we use streams and reliance locals.

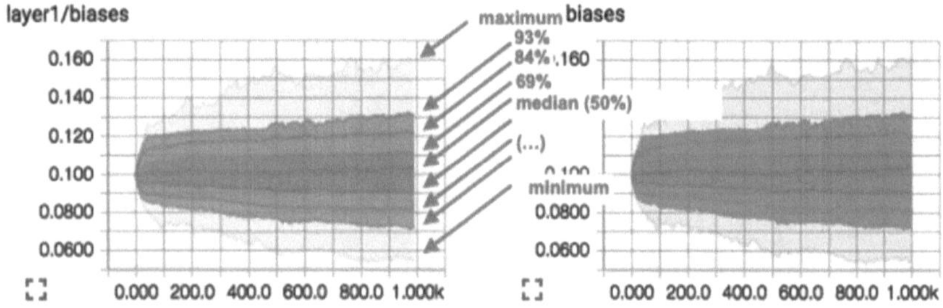

FIGURE 4.6 EEG monitoring of multithreaded CPU operations (x-axis is μs time).

4.9 FUTURE WORK

It has a few specific headings for future work. We will continue to use TensorFlow to create new and captivating AI models for man-made thought, and in the course of doing so, we may discover patterns by which the basic TensorFlow framework can be extended.[20] The open-source program can also provide a fascinating new direction for the execution of TensorFlow. Those constraints can become interchangeable elements in the execution we have designed, even through various front-end lingos for TensorFlow. Thus, a consumer can define a utility that uses Python's front finish but instead use it as a simple framework that obstructs the C++ front end from the inside. The idea is that this irritated reusability brings about an energetic system of AI pros sharing whole occurrences of their examination, yet minimal reusable segments of their exploration that can be reused in unique settings.

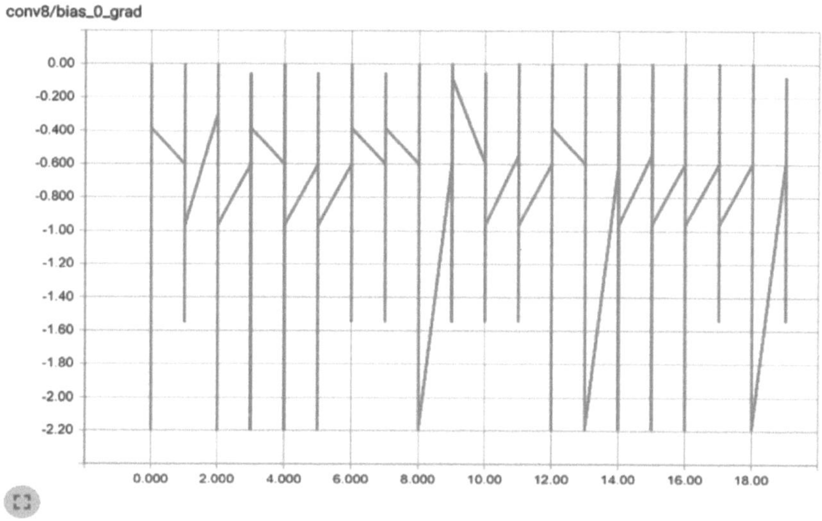

FIGURE 4.7 EEG inception training hallucination displays CPU and the GPU operation.

This compiler will understand the semantics of different turns of events, such as mixing circles, geographic blocking and tiling, specializing in unique shapes and steps, and so on. This also imagines a large area for potential research to develop the layout and center point booking computations used to figure out where various centers are going to be conducted and when they will start executing. Starting from now, we have modified various heuristics in these subsystems and we will ideally let the system figure out how to make big situational decisions.

4.10 INTERRELATED WORK

As of now, we have modified various heuristics in these subsystems and we would rather find out how to decide on major situational choices within the framework. Like TensorFlow, it promotes representative separation, makes it easier to characterize, and employs advance calculations based on gradients. TensorFlow has a C++ core that digitizes the sending models prepared in a wide variety of development environments such as mobile phones.

The TensorFlow system grants' several structure ascribes to its first framework, DistBelief,[5] and later to practically identical plans, for example, Project Adam[4] and the Parameter Server adventure. In contrast to DistBelief and scheme Adam, TensorFlow permits counts to be dispersed across different PC gadgets over various machines and permits clients to show AI models utilizing moderately elevated level portrayals. Be that as it may, the broadly valuable dataflow diagram representation in TensorFlow is progressively versatile and gradually supportive to convey a more extensive scope of AI models and improvement figuring, not in the slightest degree similar to DistBelief and Project Adam. It additionally takes into consideration a pivotal modify by requiring the assertion for crateful boundary centers as factors and part refreshing exercises in the graph that are just additional center points.

The entire separate worker subsystems gave limits to the granting and revival. The Halide structure for conveying picture taking care of pipelines utilizes a practically identical widely appealing portrayal to the information stream diagram of TensorFlow. Unlike TensorFlow, however, the Halide structure has higher level data on the semantics of its exercises and uses this information to provide particularly improved parts of data which join different undertakings, taking into account parallelism and position. Halide completes the counts on just a single PC and not in a circulated situation. In future work, we need to extend the TensorFlow with a powerful collection system for similar cross-movement.

A few other enunciated structures, including TensorFlow, were created to implement dataflow diagrams over a populace. This tells the best way of conveying about an incredible work procedure like a dataflow graph. The present nonselective helps for the control stream of subordinate information: CIEL addresses center as a significantly spreading DAG, even as Naiad uses a static outline procedure to help underscore lower idleness. Shimmer is optimized for computations that more than once acquire comparable data using "adaptable sent information indexes" (RDDs) that are sensitive state-held yields for pre-estimation. For example, Dandelion executes information stream diagrams over an exhibit of heterogeneous equipment such as GPUs as shown in Figure 4.8. The TensorFlow uses a mutt dataflow model that obtains segments from each of those structures. The information stream scheduler, the component that selects the corresponding execution core, uses a fundamental count such as Dryad, Flume, CIEL, and Spark.

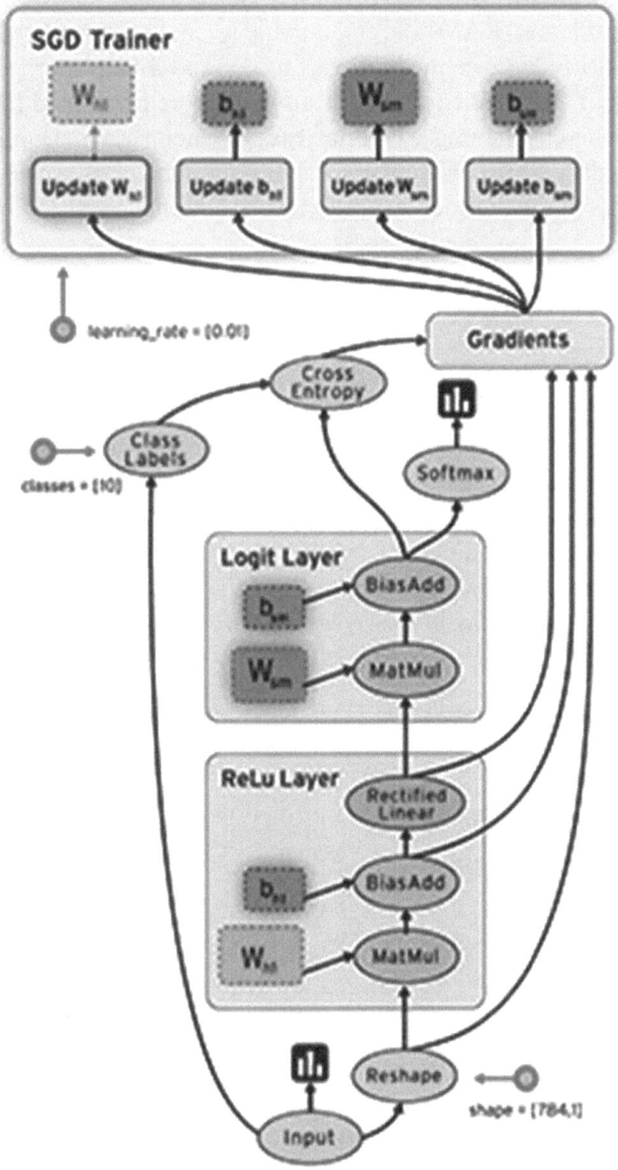

FIGURE 4.8 Timeline for single process GPU execution.

The fitting design is like Naiad, as the frame uses a solitary, dense outline of information stream to allude to the whole figure, and analyzes data on that chart on each contraction to limit bandwidth coordination. TensorFlow performs effectively, similar to Spark and Naiad, when there is sufficient RAM and in bundle to make estimate work game plans. TensorFlow accentuation uses a half-and-a-half approach: several duplicates of the identical dataflow guide may be performed for a second instantly, thereby having a similar measurement scale. Duplicates will also share data whilst also factors at the same time or use interoperability frameworks in the outline illustration work simultaneously.

4.11 CONCLUSIONS

The feature described TensorFlow, a flexible data stream-dependent programming model, almost the same as a solitary computer, and the use of this programming model is distributed. The system relies on real effort to organize research and disseminate more than 100 AIs across a wide variety of Google items and organizations. We already freely distributed a TensorFlow model and plan to create a specific network structure through the use of TensorFlow; the excitement to see how many people within Google use TensorFlow in their very own work.

KEYWORDS

- **artificial intelligence**
- **deep neural network**
- **TensorFlow**
- **heterogeneous frameworks**

REFERENCES

1. Ganesh Babu, R. Helium's Orbit Internet of Things (IoT) Space. *Int. J. Comput. Sci. Wireless Security* **2016a,** *3* (2), 123–124.

2. Ganesh Babu, R. Mismatch Correction of Analog to Digital Converter in Digital Communication Receiver. *Int. J. Adv. Res. Trends Eng, Technol.* **2016b,** *3* (19), 264–268.
3. Ganesh Babu, R. WIMAX Capacity Enhancements Introducing Full Frequency Reuse Using MIMO Techniques. *Int. J. Adv. Res. Biol. Eng. Sci. Technol.* **2016c,** *2* (16), 1–7.
4. Ganesh Babu, R.; Amudha, V. Cluster Technique Based Channel Sensing in Cognitive Radio Networks. *Int. J. Cntrl. Theory Appl.* **2016a,** *9* (5), 207–213.
5. Ganesh Babu, R.; Amudha, V. Spectrum Sensing Cluster Techniques in Cognitive Radio Networks. In *Proceedings of 4th International Conference on Recent Trends in Computer Science & Engineering, (ICRTCSE) in association with Elsevier-Procedia Computer Science*, 2016b, Vol. 87; pp 258–263.
6. Ganesh Babu, R.; Amudha, V. Allow an Useful Interference of Authenticated Secondary User in Cognitive Radio Networks. *Int. J. Pure Appl. Math.* **2018a,** *119* (16), 3341–3354.
7. Ganesh Babu, R.; Amudha, V. Comparative Analysis of Distributive Firefly Optimized Spectrum Sensing Clustering Techniques in Cognitive Radio Networks. *J. Adv. Res. Dyn. Cntrl. Syst.* **2018b,** *10*(9), 1364–1373.
8. Ganesh Babu, R.; Amudha, V. A Survey on Artificial Intelligence Techniques in Cognitive Radio Networks. In *Proceedings of 1st International Conference on Emerging Technologies in Data Mining and Information Security (IEMIS) in association with Springer Advances in Intelligent Systems and Computing Series*; Springer: Singapore, 2018c, Vol. 755; pp 99–110.
9. Ganesh Babu, R.; Amudha, V.; Karthika, P. Architectures and Protocols for Next-Generation Cognitive Networking. In *Machine Learning and Cognitive Computing for Mobile Communications and Wireless Networks*, 1st ed.; Singh, K. K., Cengiz, K., Le, D.-N., Singh, A. (Eds.); Scrivener Publishing Partnering with Wiley, 2020; pp 155–177. https://doi.org/10.1002/9781119640554.ch7.
10. Ganesh Babu, R.; Karthika, P.; Elangovan, K. Performance Analysis for Image Security Using SVM and ANN Classification Techniques. In *Proceedings of Third IEEE International Conference on Electronics, Communication and Aerospace Technology (ICECA)*, RVS Technical Campus, Coimbatore, India, 2019; pp 460–465.
11. Ganesh Babu, R.; Karthika, P.; Aravinda Rajan, V. Secure IoT Systems Using Raspberry Pi Machine Learning Artificial Intelligence. In *Proceedings of Second International Conference on Computer Networks and Inventive Communication Technologies in association with Lecture Notes on Data Engineering and Communications Technologies*; Springer: Singapore, 2020, Vol. 44; pp 797–805.
12. Karthika, P.; Vidhya Saraswathi, P. Image Security Performance Analysis for SVM and ANN Classification Techniques. *Int. J. Rec. Technol. Eng.* **2019,** *8* (4S2), 436–442.
13. Karthika, P.; Vidhya Saraswathi, P. Raspberry Pi: A Tool for Strategic Machine Learning Security Allocation in IoT. In *Empowering Artificial Intelligence through Machine Learning*; Raju, N., Rajalakshmi, M., Goyal, D., Balamurugan, S., Elngar, A., Kesawn, B. (Eds.); Apple Academic Press, CRC Press, Taylor & Francis Group, 2020; pp 133–141.
14. Karthika, P.; Ganesh Babu, R.; Nedumaran, A. Machine Learning Security Allocation in IoT. In *Proceedings of IEEE International Conference on Intelligent Computing and Control Systems (ICICCS)*; Vaigai College of Engineering: Madurai, India, 2019; pp 474–478.

15. Mekonnen, W. G.; Hailu, T. A.; Tamene, M.; Karthika, P. A Dynamic Efficient Protocol Secure for Privacy Preserving Communication Based VANET. In *Proceedings of Second International Conference on Computational Intelligence in Pattern Recognition (CIPR) in Association with Advances in Intelligent Systems and Computing*, Springer: Singapore, 2020, Vol. 1120; pp 383–393.
16. Mitola, J. *Software Radio Architecture: Object-Oriented Approaches to Wireless System Engineering*; John Wiley & Sons Ltd., 2000.
17. Nedumaran, A.; Abdul Kerim, S.; Abdu, T. S. Advanced Link State Routing Protocol Approach for Mobile Ad-Hoc Networks. *Int. J. Sci. Res. Dev.* **2017,** *5* (3), 516–519.
18. Nedumaran, A.; Ganesh Babu, R.; Kass, M. M.; Karthika, P. Machine Level Classification Using Support Vector Machine. In *AIP Conference Proceedings of International Conference on Sustainable Manufacturing, Materials and Technologies (ICSMMT 2019)*, Coimbatore, India, 2019, Vol. 2207, No. 1; pp 020013-1–020013-10.
19. Rondeau, T. W.; Bostain, C. W. *Artificial Intelligence in Wireless Communication*; Artech House: USA, 2009.
20. Tegegn Ayalew Hailu.; Nedumaran, A. A Survey on Provisioning of Quality of Service (QoS) in MANET. *Int. J. Res. Adv. Dev.* **2019,** *3* (2), 34–40.

CHAPTER 5

Introduction to Biorobotics: Part of Biomedical Signal Processing

KASHISH SRIVASTAVA* and SHILPA CHOUDHARY

G. L. Bajaj Institute of Technology and Management, Gautam Budh Nagar, Uttar Pradesh, India

**Corresponding author. E-mail: kuhusri2121@gmail.com; shilpadchoudhary@gmail.com*

ABSTRACT

Biomedical signs are perceptions of physiological exercises of living beings, ranging from gene furthermore, protein successions, to neural and cardiovascular rhythms, to tissue and organ pictures. Biomedical signal preparing targets removing huge data from biomedical signals. Biomedical signal handling includes the examination of the estimations to give helpful data whereupon clinicians can decide. It helps in finding better approaches to process the signs utilizing an assortment of scientific formula and calculations. Working with conventional bio-robotic devices, the signs can be processed by programming to give doctors continuous information and more prominent bits of knowledge to help in clinical appraisals. By utilizing increasingly modern intends to break down what our bodies are stating, we can possibly decide the condition of a patient's well-being through more non-obtrusive measures with the guide of biomedical sign preparing, scientists can find new science and doctors can screen more accurate ailments. Organisms are perplexing frameworks whose subsystems communicate, so the deliberate signs of a natural

Deep Learning in Visual Computing and Signal Processing. Krishna Kant Singh, Vibhav Kumar Sachan, Akansha Singh & Sanjeevikumar Padmanaban (Eds.)

subsystem for the most part contain the signs of different subsystems. It has countless applications and types, one of which is bio robotics. Bio robotics envelops a different cluster of orders with a heap of applications. For example, developing artificial skin that can distinguish pressure as contact is made with an object. Bio robotics may make robots that emulates or stimulates living biological organism's life more precisely or even synthetically or make living biological organisms manipulatable and practical as robots or utilize living organisms parts of robots. It is another logical and innovative territory with a remarkable interdisciplinary character, planned for expanding information on how organic frameworks work. This goal are acquired by (1) breaking down living creatures from a bio mechatronic viewpoint and (2) exploiting the received information to create inventive philosophies and advancements. Biorobotics incorporates the double utilization of a biorobots as an instrument for scholars concentrating living life forms' conduct and as proving ground in the investigation and assessment of organic examples for potential applications in engineering. As an outcome, the association between biological science and mechanical autonomy becomes two-crease: on one hand, science gives the information on the natural frameworks expected to assemble biorobots, then again, bio-roused robots speak to a supportive stage for test approval of speculations and theories planned by scientists. Biorobotics is being utilized to help in training specialists and dental specialists utilizing virtual conditions that speed the learning procedure by encouraging epiphanies. Biorobotics covers the fields of cybernetics, bionics and even genetic engineering as a collective study. Here we will be introducing some examples of soft robotics, nonorobotics having wide application of robotics in medical field while treating prosthetics and use of polyvinyl chloride (PVC), cancer fighting robots and also the next phase of biology and robotics integrations.

5.1 WHAT IS BIOROBOTICS?

The living organism contains some frameworks cooperating in a loop and customized to protect its survival. Our brain works to monitor it continuously and react to inner and outer impacts to manage internal body temperature. The pulse rate response to various actions to the sensory system, and relatively functions as a feedback system. Due to which our system uniformly releases data which gives the idea about our physical health. These information may be captured with the help of physiological devices that measures pulse, circulatory strain, oxygen immersion levels, blood glucose, nerve conduction, mind functioning, etc. Usually, such calculations and approximation clearly focuses in time by the doctors and recorded in a patient's detail chart. But approximately almost all the physicians can actually observe less than what 1% of the impact in their diagnoses. As they make prediction on the basis of what they have learned and experienced—thus the treatment choices depend on these diagnosis. This can vary for chronic diseases among the specialists. Currently, the drawback of science and medication lead to error and trial methods led by doctors. Medicines are frequently utilized in an experimentation style dependent on every doctor's encounters with their own patients whereas biomedical signal processing includes the careful study of the patient internal body behavior to give constant monitoring.

It can indicate better possibilities of chronic diseases, prior discovery of adverse events for example, respiratory failures and strokes and respiratory issue just as diseases like syncope. Biomedical signal handling is accurate and helpful in unit consideration management, where persistent information must be examined in genuine time. Bio robotics is a part of biomedical signal processing and controlling which combines as the two orders required all together first as signal and afterward producing movements. Such gadgets may likewise be utilized to specify the condition of sickness, track progress, or offer intuitive preparing encounters that can speed recuperation from a physical issue or stroke. With a comprehension of biomechanics, engineers are developing naturally propelled robots with improved and upgraded abilities over customary robots; biologically roused robots have more prominent versatility and adaptability than customary robots and regularly have tactile capacities. The world of robotics is looking to close the gaps by increasing efforts for development in these fields of medication. These advancements depend on new innovation to help in the clinical fields that were not there previously. A few organizations are utilizing 3D printing to help in the production of prosthetic. Advanced prosthetic are essential to individuals recuperating from lost appendages and constrained portability. Robotic prosthetic removes the lost appendages as well as, gives proper functioning of that body part back to the patient, yet in addition intellectually help the patient to feel healthy.

Bio robotics is the utilization of natural qualities in living beings as the information base for growing new robot plans. The term can likewise allude to the utilization of natural examples as utilitarian robot segments. Bio robotics creates robo gadgets that imitate or reenact any human being precisely mechanically or even chemically, or make biological organisms as calculative and working as machinery gadgets, or utilize organic creatures as basic requirement of these robo-gadgets. Bio robotics could utilize hereditary building to make creatures structured by artificial method.

The biomedical robotics investigating center area is revolved around the structure, improvement, and assessment of clinical level applied autonomy frameworks and savvy assistive automated stages that upgrade the physical abilities of the two patients and clinicians by means of headways in mechanical plan, displaying and control, sensors and instrumentation, figuring, and picture handling.

Core research themes right now are clinical applied autonomy, haptic interfaces, artificial intelligence, delicate applied autonomy, robot-helped medical procedure and restoration, tissue demonstrating, human augmentation, biomechanics, and human-robot interaction.

Biomedical applied autonomy inquire about naturally draws from a few controls including mechanical, biomedical, and electrical designing, intuitive figuring, applied physiology, and materials science, and is directed in close joint effort with clinical accomplices at Emory and CHOA. Key zones of utilization and interpretation incorporate input empowered automated medical procedure frameworks, robot-helped providing care; full scale mesa-miniaturized scale picture guided careful intercessions, wearable gadgets for word-related preparing and injury anticipation, and neurointegrated prosthetic gadgets. It is the utilization of natural attributes in living creatures as the information base for growing new robot structures. The term can likewise allude to the utilization of organic examples as useful robot components.

Bio robotics can create robo-gadgets that copy or recreate living natural creatures precisely or even artificially, or make organic living beings as manipulatable and useful as robots, or utilize natural living beings as parts of robots. Bio robotics could utilize hereditary building to make living beings planned by counterfeit methods. The biomedical robotics inquire about specified territory is fixated on the structure, improvement, and assessment of clinical applied autonomy frameworks and shrewd assistive automated stages that upgrade the physical capacities of the two patients and clinicians by means of headways in mechanical plan, displaying and control, sensors and instrumentation, figuring, and picture preparing. Center research themes right now have clinical mechanical autonomy, haptic interfaces, artificial intelligence, soft robotics, robot-helped medical procedure and restoration, tissue displaying, human growth, biomechanics, and human-robot connection.

Biomedical robotics research inquire about inherently draws from a few orders including mechanical, biomedical, and electrical engineering, intelligent technologies, applied physiology, and materials science, and is directed in close coordinated effort with clinical accomplices at Emory and CHOA. Key regions of use and interpretation incorporate input empowered mechanical medical procedure frameworks, robot-helped providing care; large-scale macro-mesa-micro-scale picture guided careful intercessions, wearable gadgets for word-related preparing and injury counteraction, and neurointegrated prosthetic gadgets.

These robo-gadgets in medication help by assisting clinical job pressure by regular task performance that removes their time from all the pressing duties and by creating clinical techniques better secured and a bit less exorbitant for patients. They can likewise perform precise medical procedure in small areas and exports hazardous objects. Robotic medical aides screen quiet fundamental insights and caution the medical staff at the time of need of any for treatment in the room, permitting medical workers to examine some patients immediately. These robotic assists more carefully as a result of which it enters data into the patient's e-health record. Automated vehicle is sometimes found to be traveling through medical clinic, halls conveying supplies. Robots are likewise helping in medical procedure, permitting the senior medical staff to lead medical procedure through as small cut instead of an inches-in length entry point. Biorobotics is consisting of a huge effect in various territories of medication, too. Robotic technologies show up in different aspects that legitimately influence quiet consideration. These gadgets may be used to sanitize persistent rooms and working suites, decreasing health hazards for patients and clinical staff.

These robots are appointed at research centers to experiment and to ship, examine, and save them. We have experienced that it takes a lot of time and proper examination and experience for a person to locate an "apt vein" to draw blood from it. The automated labs can get that vein and draw the blood fluid with less pain and anxiety for the patient. Robots are likewise to get ready and dispense medication in pharmacological labs. There are several applications of bio robotics in the medical field that can help in treating the patient at a high pace and some of these applications are mentioned below:

- **Telepresence:** Telepresence physicians use robots to assist them with looking after and treatment of the patients in rustic or undeveloped regions, providing "Telepresence" inside the room. "Masters can be accessible as needs to be, in the form of robot, to respond to doubts and lead the treatment from undeveloped regions."
- **Surgical assistants:** These button-controlled robot gadgets assist doctors in performing tasks, generally insignificant invasive techniques.
- **Rehabilitation robots:** These assume a pivotal work in the recuperation of a single person with incapacities, with better versatility, quality, coordination, and personal satisfaction. These gadgets can be improved to vary the status of every patient's health as they recoup from strokes, horrible cerebrum or spinal string wounds, or neurobehavioral or neuromuscular sicknesses, for example, numerous sclerosis. Augmented reality incorporated with restoration robots can likewise improve equalization, strolling, and other engine capacities.

Bio robotics is a huge area under biomedical signal processing and controlling section comprises of topics such as biomimetic, bio brick, humanoid robot, etc. which is been explained in details further in this book.

5.2 BIOMIMETIC

Definition: Biomimetic also known as biomimicry is the impersonation of the structure, frameworks, and components of the environment to tackle complex human issues.[1] Biomimetic has offered ascend to new advances propelled by natural arrangements at full scale and nanoscales.

Biomimetic are on a fundamental level to be applied in numerous fields. As a result of the assorted variety and unpredictability of organic frameworks, the quantities of highlights that may be copied are enormous. Biomimetic usages are at different phases of advancement from advances that may turn out to be economically usable to models.[2]

5.3 BIOPRICK

Definition: Bioprick parts are DNA successions which adjust to a limitation protein get together standard.[3,4] These structure units are utilized to create and combine bigger nonnatural biological track by singular parts and mixes of parts with characterized capacities that is later combined with the living cells, for example, Escherichia coli cells to develop new organic systems. Examples of bio brick parts incorporate advertisers, ribosomal restricting locales (RBS), coding groupings, and eliminators.

5.4 HUMANOID ROBOTS

Definition: A humanoid robot is a gadget with its model figured as such to take to look alike to a normal human being. These models might be for practical purposes, for example, interfacing with human apparatuses and conditions, for exploratory purposes, for example, the research over bipedal velocity, or for various reasons or purposes. While all is done as to be told, then humanoid robots is in the middle, designed same as the human body shape with a head, two legs and hands. Many of such humanoid robots may demonstrate only one part of the body, for example, from the abdomen up.

A few of these humanoid robots additionally creates a beeline for imitate human facial highlights, for example, eyes and mouths. Androids are humanoid robots designed to take the place or as replacements over human beings.

5.5 CYBERNETICS AND ITS APPLICATION

Cybernetics is appropriate when a structure is being broke down and fuses into a closed flagging loop—at first indirectly to as a "roundabout causal" relationship—that is, there are actions taken place by the structure creating few modification in its state and that makes the difference is reflected in the structure in few ways (input) which triggers a system change (Fig. 5.1).

The basic objective of the wide field of robotics is to comprehend and characterize the capacities and procedures of structure that have objectives and that take an interest in roundabout, and these chains transfers from motion to detecting to correlation with wanted objective, and returns to motion. Its center is the procedure by what anything (digitalized, mechanical, or natural) produces information, act according to data, and makes the difference or can create differences to all the more likely achieve the initial two errands.[5] Cybernetics in science is the investigation of robotic framework present in natural life forms, principally concentrating on how creatures adjust to their condition, and how data as qualities is passed from age to age. There is additionally an auxiliary spotlight on joining counterfeit frameworks with natural systems.[6]

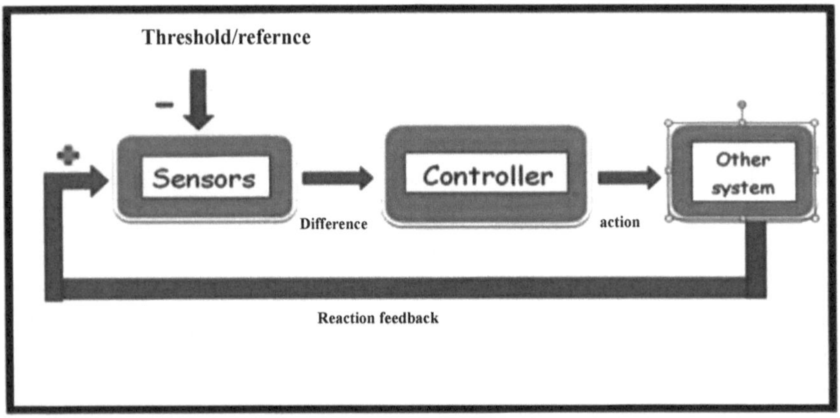

FIGURE 5.1 The cybernetics cycle.

Cybernetics is an order which reviews correlation also controls the human being and robotics worked by man. An increasingly theoretical explanation, recommended by Louis Couffignal in 1958, which says robotics are "the pillar for guaranteed productivity for a task." The word artificial intelligence was rethought by Norbert Wiener in 1948 (the Greek cabernets, pilot, or rudder). The word was first used by Plato in the feeling of "the specialty of directing" or "the craft of government." Ampère utilized the word artificial intelligence that represented "the study if procedure for overseeing." The first ever cybernetic system presented was to control the speed of motor (steam), developed by James Watt and Matthew Bolton in 1788, he was declared as a representor, or as a controller of ball. Digital science has in reality a similar concept as government: the uniqueness of supervising and correlating exceptionally complex frameworks.[7]

5.6 ORIGIN OF CYBERNETICS

The complete historical revolutions where partitioned into four extensive stretches that we indicated as four creation standards: Hunter-Gatherer; Craft-Agrarian; Trade-Industrial; and Scientific-Cybernetic.[8] Every one of these creation standards begin with an incredible technological leap forward, which we mean as a production insurgency. There were three described as follows:

- The Agrarian or Neolithic Revolution
- The Industrial Revolution
- The Cybernetic Revolution (developing)

5.6.1 THE CYBERNETIC REVOLUTION

The Cybernetic Revolution is an extraordinary leap forward from technological creation to the production and administrations dependent on automatic system. Its underlying stage goes back to the 1950–1990s. The leaps forward happened in the circles of mechanization, vitality creation, creation of manufactured materials, space advances, investigation of room and ocean, horticulture, and particularly for the improvement of digital device facilities, correspondence, and data. We signify this underlying period of the Cybernetic Revolution as a logical data also the last phase of period of self-regulating system.

As for now, we are in an upgraded stage that would presumably go on until the 2030s (Fig. 5.2). The moderate stage is a time of quick circulation and modification in the developments created at the past stage (e.g., PCs, Internet, phones, and so on.). The digital and living way conditions are additionally arranged to be beneficial for future to leap forward. It can be predicted that the last period of the Cybernetic Revolution will prompt the rise of different self-regulating system for a time duration of around 2030–2070.[8]

While in the past much cybernetics experiments were depended on hypothesis and theory, nowadays there are hard instances of artificial intelligence surrounding us that address technological, physical, organic, psychological, and social structure. Future medical technologies having been thinking about over the things like constant well-being checking like a self-regulation system. For the time interval of last period of the Cybernetic Revolution, a significant heading of self-guideline might be executed utilizing biosensors, information correspondence, and health observing frameworks for early conclusion and sickness anticipation. One can envision that later on, such innovation will turn into a necessary piece of a living being's life, giving a consistent scanner of a life form or a specific body part and transfers the data to an emergency clinic if there should arise an occurrence of potential or genuine hazard. All in all, medication will create toward expanding individualization and representation by methods for singular treatment, while the utilization of mass medications and standard remedial innovations will be diminished.[8]

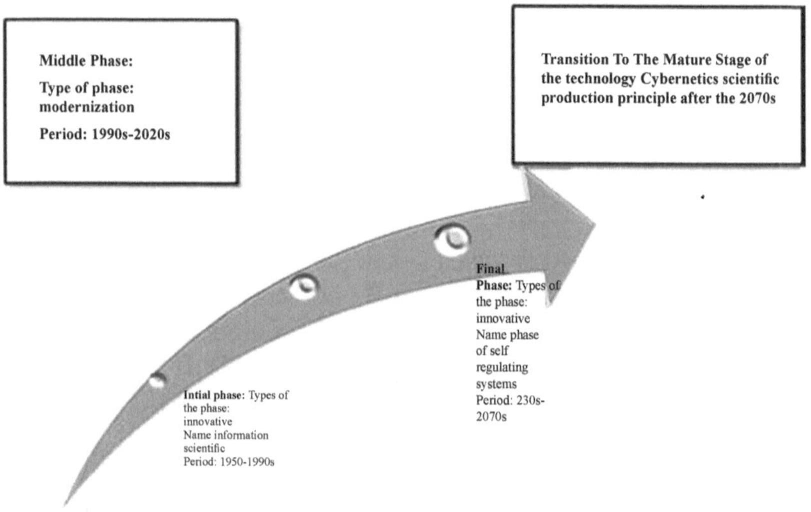

FIGURE 5.2 The three phases of cybernetics.

Some of the applications based on cybernetics that are developed or are on the process of developing in this era of 21st century have been stated here.

5.6.1.1 FULL-SPECTRUM VISION

The manner in which people can "see" in our general surroundings has been changed a great deal in the last century. There are presently telescopes that can see significant stretches through space and focal points so little they could transform your eyes into cameras. Human visual perception itself appears to be entirely restricted contrasted with these models, yet that is all on the edge of progress. Researchers from the University of Massachusetts have built up an approach to give night vision to mice and maybe one day humans. The mice were given the superpower of having the option to find in obscurity with only a straightforward infusion containing nanoantenna. The leap forward from implies later on people could have night vision at whatever point they required it.

5.6.1.2 BRAIN-COMPUTER INTERFACING

Our brains are unfathomably incredibly unique organ. Long periods of examination into the manners in which our psyches work is paying off as innovation likewise quickly improves. Scientists from Cornell University have professed to have imagined the world's first non-obtrusive mind to cerebrum interface called BrainNet. The interface joins electroencephalography (EEG) to record brain signals and transcranial attractive incitement (TMS) to convey data to the mind. EEG explore that helps individuals to figure out how to control and comprehend their own EEG can possibly help beat mental clutters and increment mind perception.

5.6.1.3 PSYCHOKINESIS

Psychokinesis or the capacity to move objects with your brain did not depend on science. However, there are a lot of new advances in technological ideas working on it artificially. There are no psychics, and there have never been. However, that doesn't imply that we can't make techno-psychics artificially. Specialists are dealing with strategies for remote control of automatons utilizing brain-computer interface tech. One such group can be found at the University of Minnesota that permits clients to remote control quadcopter rambles with their considerations. The controller wears a uniquely planned top with cathodes that transmits the little electrical signs to a PC. Expert programming at that point changes over the electrical signs into guidelines to move the automaton remotely. Volunteers should initially adjust to this peculiar strategy for control; however, can get sufficiently proficient to explore the automaton through a deterrent course.

5.6.1.4 DISEASE IMMUNITY

Researchers are dealing with approaches to detect diseases and prevent it before hand from our bodies. These nanorobots could be infused into our circulatory system where they decreases the chances of becoming ill and even gives health report information to your cell phone by means of the cloud. A striking use to the science universe will be, in 1955, the physicist George Gamow distributed a farsighted his views on Scientific American called "Data transfer in the living cell," and artificial intelligence gave researcher Jacques Monod and François Jacob a combination of sentences for defining their initial hypothesis of quality administrative systems in the 1960s. This is further class field as biocybernetics.

Biocybernetics is simply another naming plan (the expression "robotics" itself started as a reflection about natural frameworks working) utilized in artificial intelligence as a depiction of organic science comprehended in innovative terms, made out of natural trains that profit by the use of cybernetics including nervous system science and multicellular frameworks. Biocybernetics assumes a significant job in system biology, trying to coordinate various degrees of data to see how natural frameworks work. And then there is another branch of cybernetics that plays an important role in developing new technology and its advancement in the field of biomedical signal processing and controlling that is medical cybernetics. Here we will be explaining the actual meaning of medical cybernetics and its specification which can lead to further innovative development in the field of biomedical signal processing and controlling.

Medical cybernetics is a part of artificial intelligence, influenced by the improvement of the technology, which applies the ideas of cybernetics to clinical research and practice. It covers a rising working project for the utilization of structural system and correspondence hypothesis and choice hypothesis on biomedical research and health-related inquiries. Medical cybernetics look for quantitative depictions of natural elements, it explores intercostal arranges in human body biology, clinical dynamic, and data preparing structures in the living life form.

5.6.1.5 TOPICS IN MEDICAL CYBERNETICS

- Systems hypothesis in medical cybernetics: The extent of system hypothesis in the medical cybernetics is looking for and demonstrating of physiological elements in the unblemished and sick life form to increase further bits of knowledge into the authoritative standards of life and its perturbation.

- Medical information and communication theory: Motivated by the familiarity with data as a basic guideline of life the use of correspondence hypothesis to biomedicine intends to numerically portray flagging procedures and data stockpiling in various physiological layers.

5.7 WHAT IS NANOTECHNOLOGY?

Nanorobotics is a developing innovation field making gadgets or devices whose segments are at or approach the size of a nanometer (10–9 m).[9–11] As presented by the National Nanotechnology Initiative, it has a complete access over objects at the nanoscale.[12] They are the gadgets that permit exact connections with nanoscale materials, also may control nanoscale goals. The nanorobotics field is completely not the same as that of macroscale robots based on estimation and the materials utilized; however, there are a few commonalities among the structure and techniques of controlling.

The field of clinical nanorobotics has accomplished extensive advances. Though, a few issues and difficulties must be tended to before macroscale/nanorobots might contain true medical applications. The objective of the medical model isn't just to assess the helpful productivity of the stages; however, to recognize the clinical hazard, as assessing askew impacts of nanorobots is as basic as assessing viability. Undoubtedly, they may have a lot of difference among the goals of clinical nanorobotics and today's reality, as the heritage of science fiction has defined the reasonable limits of what's in store well before researchers could. The assembling of the macroscale/nanostructure motors must be enhanced, with extraordinary thought for compatibility of biological objects and corruption, to highlight in vivo safety concerns.[13]

Furthermore, appropriate models ought to be built up to explain the benefits of macroscale/nanorobot treatments over conventional techniques which as of now satisfy FDA guidelines. In any case, macroscale/nanorobotics may possibly improve clinical determination and treatment. We ought to likewise consider that the structures and goals of a little gathering of researcher and engineering could before long influence the lives of a great many individuals directly and significantly; in this manner it becomes fundamental to make monetary as the prior of all, social and moral ramifications of the utilization of clinical nanorobotics. These suggestions are probably going to be comparable among all those most critical innovative upsets.[13]

Application of nanobots in biomedical designing is significant most definitely. Nanobots can end up being the modest, plausible, and snappy arrangement that the world has been looking and has evaded us for quite a while. Biomedical nanobots additionally have an extremely high achievement pace of more than 80% in the entirety of their application and the way that solid cells aren't influenced by any medication activity in treatment of malignant growth can help in drawing out the life of the patient. The present patterns show incredible guarantees in clinical nanobots. The applications develop at an exponential rate each year and the attainability of every one of these applications additionally increment. Some future uses of nanobots can incorporate preemptive location of deadly diseases like cancer, diabetes, Alzheimer's, and so forth, medical procedure of the sensitive heart muscles, further developed mapping of the mind and even increment life span of life. Nanorobotics is a developing field and gives gigantic guarantee in taking care of numerous issues which were viewed as a, immovable object before people and this is only the start of another unrest in the clinical field and soon enough, it won't be an exaggeration to state that we will be godlike one day.[13]

The expression "nanotechnology" introduced first was utilized in detail in 1974 to allude to the inexorably exact mechanizing and designing an instrument, advancing from bigger to littler scopes and eventually to nanoscale resistances, going through the procedure as Feynman's proposed "top-down" approach, all through the 1980s.[14,15]

As indicated by Richard Feynman, that was his previous graduate student and colleague Albert Hibbs the one that initiated the proposal to him (around 1959) the possibility of a clinical use for Feynman's hypothetical miniaturized micro machines (see natural machine). Hibbs proposed that specific machines may somehow be decreased in smaller size to a level that it could be one day, in principle, be conceivable to (as Feynman puts it) "swallow the specialist." The thought was fused into Feynman's 1959 paper "There's Plenty of Room at the Bottom."

A developed nanomedicine may need capacity to construct its shape and gadgets to nuclear precision, subsequently atomic nanotechnology and subatomic assembling are key empowering innovations for nanomedicine. Feynman proposed utilizing machine instruments to produce tiny machine apparatuses, these to be utilized thus to produce even tiniest machine devices, etc. right deep into the nuclear level. Feynman prophetically inferred this is "an advancement that we believe that can't be kept away from." Such nanomachine apparatuses, like robots and gadgets at nano-level could at last be utilized to build up a wide scope of submicron device designing and assembling devices, that is, nanotechnology. Feynman's proposed use of these devices included delivering immense amounts of ultra small PCs and different small scale and nanorobots.[14]

Feynman was unmistakably mindful of its capacity for clinical utilizations of the new ideas he was proposing. In the wake of talking about his thoughts among his partner, Feynman gave a proposal to let the main known proposition for a nanomedical method to treat cardiac diseases

5.7.1 NANOMEDICINE: BIRTH OF NEW ERA

Conclusive confirmation about presence of smaller atomic units was not evolved till the end of the 19th century. This might clarify the reason for possibility of nanomedicine is an only 20th century wonder. The principal trace of it comes to existences through the celebrated 1929 article composed by J. D. Bernal. "The revelations of the 20th century, especially the nanomechanics of the Quantum Theory which addresses the idea of issue itself, are undeniably increasingly principal and should on time lead unquestionably progressively significant outcomes. The initial measurements are the advancement of new objects and new procedures where physics, chemistry, and mechanics will be inseparably intertwined. The level ought to before long be arrived at when the objects could be delivered that are not only difference to what our environment has provided in the form for stones, metals, woods, and filaments which brought light over the presences of atomic engineering.

As of today, we have an idea about all the assortments of molecules; that are starting to know about the power of that of dilemma it as one; soon we will do this in an approach that gives us motivations. The outcome—not all that extremely far off—will likely result into death of the period of metals and everything that it suggests—mines, heaters, and motors of huge development. Rather we ought to have a universe of texture products, light and versatile, solid just for the reasons it is always been utilized a society which will impersonate the fair flawlessness of an organisms. The revelations of the 20th century, especially the nanomechanics of the Quantum

Theory which address the idea of issue itself, are undeniably increasingly principal and should on time lead unquestionably progressively significant outcomes.

The initial measurements are the advancement of new objects and new procedures where physics, chemistry, and mechanics will be inseparably intertwined. The level ought to before long be arrived at when the objects could be delivered that are not only differences to what our environment has provided in the form for stones, metals, woods, and filaments which brought light over the presences of atomic engineering. As of today, we have an idea about all the assortments of molecules that are starting to know about the power of that of dilemma it as one; soon we will do this in an approach that gives us motivations. The outcome—not all that extremely far off—will likely result into death of the period of metals and everything that it suggests—mines, heaters, and motors of huge development. Rather we ought to have a universe of texture products, light and versatile, solid just for the reasons it is always been utilized a society which will impersonate the fair flawlessness of an organisms.

Improvement of the idea of nanomedicine has two paths that it follows like the ways which Richard Smalley has named "wet nanotechnology" in the natural convention, and "dry nanotechnology" in the mechanical custom. The two methodologies were foretold in theoretical fiction. The accompanying truncated history centers around nanomedicine to a great extent to the rejection of more extensive issues in atomic building, producing and nanoscopy, and incorporates various motivational, theoretical, or anecdotal early references from non-refereed sources.

The logical custom of natural nanorobotics for clinical purposes started in 1964 when Robert Ettinger, a cryonics pioneer, proposed that cell level as well as atomic level fix may be created forever expansion. Ettinger guessed that "...surgeon machines, functioning each hour of the day in day-to-day basis for quite a long time or even hundreds of years, will softly re-establish the solidified mind{brain function}, cell by cell, or even particle by atom in basic regions."

In 1972, Ettinger proposed utilizing hereditary building to make tiny biorobots: "genetic engineering most sensational effect which highlights requirement of the adjustment of people; however it will have different uses also. A portion of the 'robots' that are going to work for us should be nanominiaturized." Existing living beings could be changed to make organically trained biorobots for clinical approach: "In the event that we can structure adequately complex personal conduct standards into infinitesimally little organism, there are evident and unlimited prospects, probably the most significant in the clinical territory. Maybe we can convey gatekeeper and scrounger creatures into the blood, better than the leukocytes and different operators of living organism legacy, which will productively chase down and clear out a wide assortment of unfriendly or harming trespassers." Computerized cell fix machines "must utilize means, for example, protein blend and metabolic pathways to analyze and fix any wastage of data and can be saved by determining the fact that fix programs consolidate fitting RNA tapes into itself."

Additionally in 1972, Danielli depicted different opportunities for producing new life structures through "life-combination" and hereditary building, noticing that "macromolecular designing" may empower the improvement of exceptionally amazing and minimal macromolecular PC frameworks. Portrayed different opportunities for producing new life structures by means of "life-combination" and hereditary building, taking note of that "macromolecular engineering" may empower the advancement of ground-breaking and smaller macromolecular PC frameworks.

In 1970, Volkenstein noticed that "the production of a nonmacromolecular framework that could go about like a shape for living organism forms is certainly conceivable" yet couldn't emerge itself, and that the macromolecularity of present organism isn't fundamental; however, because of their transformative sources. Collecting from some poetic permit, he included: "Thus, the computerized nonmacromolecular gadgets, which stimulates life, might have been and may be made on Earth just by man. At that point it could consummate itself unbounded."[16,17]

In 1981, Drexler recommended the development of precisely deterministic nanodevices utilizing organic portions; these gadgets could review cells at the atomic level and furthermore fix cell tissues which can be harmed while cryonic suspension. In 1982, Drexler portrayed cell fix devices significantly much more plainly in the mechanical custom in a famous production.[18,19] By 1983, Drexler started alone writing a number of documents for the specialized paper entitled as "Cell Repair Machines" which reinitiated the research, and gave us information, regardless of whether a progressed mechanical-based nanotechnology would grant development of frameworks of subatomic scale sensors, PCs, and controllers ready to go inside and fix cells; the behavior of the mathematical calculations and size of the robotics parts required to get assembled are assets expected to manage fixes; and the mechanical abilities and limitations critical to the fix procedure (in request to) get the idea and outline of theoretical structure of a cell fix framework dependent on a develop subatomic innovation.

The advancement of nanorobots is achieved by utilizing different methodologies as discussed below.

5.7.1.1 BIOCHIP

The mixture of nanotechnology, photograph lithography, and new biomaterials, may look like potential route initial for structuring innovation to create robo-gadgets (of nanometer range) for clinical approaches, for example, diagnosis and medication conveyance. This reasonable methodology in structuring nanorobots is an approach which is utilized in the electronic enterprises.

5.7.1.2 POSITIONAL NANOFACTORY ASSEMBLY

By the year 2000, Robert Frietas and Ralph Merkle discovered nanofactory cooperation that is a progressing exertion comprising of 10 associations with 23 analysts from 4 nations. This cooperation targets growing positionally controlled mechanosynthesis and diamondoid nanofactory that can fit for building a diamondoid clinical nanorobot.

5.7.1.3 BIOHYBRID NANOROBOTS

This classification of nanobots includes the amalgamation of organic particles and electronically created waves empowering fast development of the bot to different areas relying upon the prerequisite. Biohybrid frameworks coordinate engineered structures in the range of nanometer which can move spontaneously as the engine of the small scale/nanorobot.

These nanotechnology gadgets will significantly affect the lives of amputees: Envision if the gadget was shrewd well fix the prosthesis parameters by themselves to get in any circumstance where we cooperate with the Earth—conveying various measures of burden, strolling on sand or grass—and the amount more amputees may have the option to depend on their prosthesis in their regular day-to-day existence. This is the following phase of worl.[20]

5.8 APPLICATION OF NANOROBOTICS TO VARIOUS DIMENSIONS OF MEDICAL AND HEALTH CARE FIELD

The flexibility and ease provided by nanobots guarantee us that there can be regions in medication and other related fields that can be incredibly be profited by this development. In light of the utilization, the bionanorobots can additionally be categorized into three classes as follows:

- Nanobots for discovery
- Nanobots for analysis
- Nanobots for treatment

5.8.1 DIAGNOSIS AND TESTING

Clinical nanorobots are used as to target for diagnosis, testing, and checking of viruses and bacteria, etc. as well as tissues and cells into the blood system. These robo-gadgets (of nanometer in range) are selected for observing and recording, and report some crucial signs if any, like for an example, temperature, weight, and invulnerable framework's parameters of various pieces of the human body constantly. So, we are now in era of developing new innovative ideas and methods to overcome several inabilities faced by the people in the world.

5.8.2 PROSTHETICS AND NANOROBOTICS

For example, Minoru Hashimoto, a teacher at the Shinshu University in Japan, has planned a wearable robot to help an individual's hip joint while strolling. These frameworks are a plasticized polyvinyl chloride (PVC) gel with work terminals and provided with some current contracts like muscle. Propelled prosthetics are basic to individuals recuperating from lost appendages and constrained portability. Robotics prosthetics not just supplant the lost appendages, returning its usefulness to the patient, yet additionally intellectually which contributes in making the patient feel better at mental level.

The following period of nanotechnology and human reconciliation is to reproduce human organs and working with robotics. A wearable robot helps an individual's hip joint while strolling. The wearable framework is a PVC gel with work terminals and provided with current. The work anodes making layers of the gel and as the current is provided, the gel flexes and contracts simply like a muscle would. Basically, the robotics framework is a wearable actuator that makes development[20] represents the wearable robot to help an individual whose hip joint functions normally while strolling.

5.8.3 TUMOR TARGETING DNA AND ROBOTICS

Robots are likewise not supporting the outside of the human body yet in addition inside too. Robotic development is presently happening at the nanoscale. Nanomedicine is a part of medication that joins nanotechnology for creating infinitesimal, particle estimated as nano-sized particle to analyze and treat troublesome ailments like malignant growth. The nanostructures can overlay themselves into a wide range of shapes in sizes at a size of one thousand times littler than the width of a human hair. "We have built up the first completely self-ruling, DNA automated framework for an exact medication plan and focused on malignancy treatment." Also, this innovation is a technique that can be utilized for some kinds of malignant growth, since all strong tumors taking care of veins are basically the equivalent.[20]

The arrangements were made to utilize these nanofigures to remove the blood supply by prompting blood coagulation with better remedial viability and well-being profiles in various strong tumors utilizing DNA-based nanocarriers. Yan overhauled the nanomedicine to be a self-ruling robotics framework, one ready to finish the task all alone. Every single nanorobot is produced using a level, rectangular DNA origami sheet of 90 nanometers by 60 nanometers in size and consist of blood-coagulating compound known thrombin. Thrombin can square tumor blood stream by thickening the blood inside the tube that feed tumor development, prompting tumor tissue demise.[20]

Approximately out of all four thrombin atoms were connected to a level DNA framework, and the sheet was collapsed in on itself like a piece of paper into a hover to make an empty cylinder. A unique payload known as a DNA aptamer was incorporated to guarantee that solitary the particular protein nucleoli was focused on so just the malignant growth cell was attacked. Once bounded with the tumor vein surface, the nanorobots convey the medication to the tumor's blood vessel and the thrombin starts coagulating blood. Within 24 h, the tumor tissue starts getting harmed, and following 14 days of treatment, the tumor starts to contract.[20]

5.8.4 NANOROBOTICS IN GENE THERAPY

Nanorobots are additional materials that are or can be utilized for the treatment of hereditary ailments, by connecting the structure of DNA and proteins inside the atom. The adjustments and abnormalities inside DNA and protein successions that are remedied (treated). The chromosomal replacement treatment is productive inside the cell to fix. An amassed fix vessel is fixed and found inside the human body to support genetic qualities by skimming deep in the bottom of the cell. Supercoil of DNA at time at which it is expanded into the lower pair of robotic arms, the nano-machine grabs the strand which is unfolded from studies and diagnosis; in the interim the upper arms isolate the proteins in the chain. The details that are kept inside the enormous nanocomputer's informative file is placed externally from the core and contrasted and the nuclear structures of both DNA and proteins that are related to the correspondence interlinked cell fixing transporters. Irregularities discovered in the shapes are rectified, and the proteins are matched to the Deoxy Nucleic Acid chain indeed changes inside their unique structure.

5.8.5 ADVANCEMENT IN SURGERIES BY USING NANOTECHNOLOGY

Deoxy Nucleic Acid chain indeed changes inside their unique structure. Surgical nanorobots are placed into the vascular system in a human body and various parts. These nanorobots go about as semiautonomous site nearby white blood cell into the human vascular system and then changed as well as coordinated by one of the protective cells present in human body.

This modified surgical nanorobot can do a variety of jobs like scanning for pathogens, and afterward diagnosis and small changes in the injuries in the treatment process by nanocontrol synchronized by a system while monitoring and entering with supervisory specialist through coded ultra-sound signals.

Recently, nanomedical procedure is being studied. For example, a micropipette quickly vibrating at a recurrence of 100 Hz micropipette similarly below 1 micron tip the difference of this distance is helpful in cutting dendrites from single neurons. This method doesn't tend to harm the cell capacity.

5.8.6 CANCER DETECTION, TREATMENT, AND NANOTECHNOLOGY

The present phases of clinical advances and treatment instruments are utilized for the effective measurement for fighting against cancer. The significant viewpoint to get the best treatment results it all depends in the improvement of proficient medication conveyance to diminish the reactions from the chemotherapy.

Self-impelling attractive nanorobots equipped for natural route in biological fluid with improved pharmacokinetics and more profound tissue infiltration involves promising technique in focused cancer growth treatment. Further, nanorobot reduces ex vivo HCT116 tumor spheroids more effectively than free DOX. The multifunctioning nanobots plan speaks to a progressively articulated technique in focusing on tumors with self-helped anticancer medication conveyance for "broad" locales in treating diseases. Nanorobots with installed synthetic biosensors are helpful and used for identifying the tumor cells at the initial stage of the disease improvement into a patient's body. Nanosensors are likewise used to discover the force of E-cadherin signals. Though, creating nanorobots for biological usefulness is yet an experiment as it takes some innate constraints, for example, complex arrangement innovation, trouble of surface alteration, trouble of movement in biological fluid, and relying upon the material, poor biocompatibility or biodegradability.[21,22]

New DNA Nanorobots Target and Destroy Breast Cancer Cells:

Researchers have distinguished a little succession of DNA (aptamer) that can perceive and tie to HER2 proteins and focusing on them to the lysosomes. But this aptamer is less steady in the blood (serum), so that, the researchers looked to know whether including another DNA nanostructure (tetrahedral system nucleic corrosive (tFNA)) could upgrade the biostability and the counter cancer growth properties of the aptamer, to figure this out, the investigation collaboration has planned new DNA nanorobots which comprise of (tFNA + HER2 aptamer), and when these nanorobots were infused into the mice, it continued in the blood for quite a while about more than the twice the length of the aptamer alone.

From that point onward, the analysts tried the nanorobots on three bosom malignancy cell lines in vitro, and it indicated that these nanorobots executed the HER2 positive disease cells as it were. The outcomes additionally demonstrated that the tFNA expands the pace of official of the aptamer to HER2, in this manner, decreasing the degree of HER2 on the cell surface.

5.8.7 NANODENTISTRY

Nanodentistry is among the highest usage field as compared to the rest because nanorobots help in various procedures engaged with dentistry.

Nanodentistry is among the highest usage field as compared to the rest because nanorobots help in various procedures engaged with dentistry. These gadgets are utilized in desensitizing tooth, oral sedation, fixing of sporadic positioning of teeth in right order and treatment for better teeth strength, significant tooth fixes and improvement of outlook of teeth, and so forth. The utilization of nanotechnology to dentistry also the amount time needed for actualize the after effects of examination in training are the principal addresses that emerge with respect to nanotechnology in dentistry.

Like nanomedicine, the improvement of nanodentistry will permit almost consummate oral well-being by the utilization of nanoproducts and biotechnologies, along with tissue building and robots of nanometer ranges.[23]

Nanocrystalline technology can be used for modifying these bone rafts. Moreover, it demonstrated that nanocrystalline hydroxyapatite invigorated the cell multiplication essential for periodontal tissue recovery. Although clinical robots are not foreseen to affect dentistry sooner rather than later, it isn't too soon to consider their potential effects.[32]

Dental nanorobots can travel through teeth and encompassing tissues by utilizing explicit development instruments. Nanocomputers that have been recently modified by means of acoustic signs utilized for ultrasonography can control nanorobotics functions.[23] Developments in computerized dental imaging strategies are likewise expected with nanotechnology. In advanced radiographies acquired by nanophosphor scintillators, the radiation portion is decreased and great pictures are gotten.[38] It is beneficial at the time of surgeries in dentistry as well.

5.8.8 TISSUE ENGINEERING AND DENTISTRY

Potential utilizations of tissue engineering and undifferentiated organism inquire about in dentistry incorporate the treatment of orofacial fracture, bone enlargement, ligament recovery of the temporomandibular joint, pulp repair, periodontal tendon recovery, and embed osseointegration. It empowers placing of the implants that wiped out a delayed recuperation period, are biologically and physiologically more steady than recently utilized embeds, and can securely bolster early stacking.[24,25]

Studies identified with the recovery of bone tissue establish a significant piece of the investigations in this field. Nanoscale filaments are comparable fit as per the arrangements among collagen fibrils and hydroxyapatite gets fixed in skeleton system. The polymers that doesn't harm our environment or earthenware ingredients which are favored in bone tissue engineering might not have adequate mechanical perseverance regardless of their osteoconductive and biocompatible characteristics even after their osteoconductive and biocompatible characteristics.

Theories acted as of late demonstrate that these particles of nanometer range can be utilized to upgrade the properties of these materials. The fundamental purpose behind inclining toward nanoparticles is that the scope of measurement of these structures is equivalent to that of cell and atomic parts.[26,27]

Challenges and difficulties faced by nanotechnology in the medical field are as follows:

- Exact situating and assembling of nanoscale parts.
- Low budgeting nanorobot mass assembling techniques.
- Synchronization of various free nanorobots.
- Biocompatibility concern.
- Monetary and strategic concerns.
- Deficient absorption for treatment measures.

Nanotechnology will carry tremendous modifications in medication and dentistry field. Although, with advancements in the field, it might likewise represent a hazard for abuse and misuse as well. Time, more up-to-date improvements, practical and specialized assets, and our requirements will be figured out about the uses to be acknowledged first.

5.9 APPLICATION OF NANOTECHNOLOGY

5.9.1 MANAGEMENT OF DIABETES MELLITUS

Detection of insulin and blood sugar level using nanotechnology advancement: Another strategy which utilizes nanotechnology to observe and calculate with least time required. It measures insulin and blood sugar level is a significant advance toward building up the capacity to calculate the strength of the body's insulin-producing cells. It very well may be accomplished in the given manner.

5.9.2 MICROPHYSIOMETER

The microphysiometer is worked from multiwalled carbon nanotubes, which resemble a few level layers of carbon molecules kept one over the other and folded to form tiny tubes of about nanometer. The nanotubes are conductive in nature and the convergence of insulin inside the vessel and it can be identified easily with the flow at the anode and the nanotubes work dependably at pH levels are the properties of living cell.

Recent work on identification techniques to measure insulin production at interims by collecting it in a set time intervals is a little example and estimating their insulin levels. The advanced sensor distinguishes insulin levels uniformly by estimating the exchange of electrons are released when insulin particles oxidize in the presence of glucose. At the point when the hormones produces more insulin atoms. The applied voltage at the sensor increases and in the other way around, permitting checking insulin concentrations progressively.[28,29]

5.9.3 IMPLANTABLE SENSORS

Utilization of polyethylene glycol globules covered with fluorescent particles is infused below the skin layer and stays in the liquid present in body cell. Exactly at time when glucose is in the liquid present in the cell of the body drops to the hazardous levels, glucose takes the place of fluorescent atoms and makes aglow. This gleam gets visible on the tattoo set over the skin. Sensor microchips are likewise being grown to continuously screen initial or basic body report including pulse, temperature, and blood glucose. A biochip can be embedded under the skin and transfers a signal that should be observed consistently.[29]

5.9.4 DEVELOPMENT OF ORAL INSULIN

It has gotten increasingly possible for the creation of insulin through nanotechnology. At the point when insulin is given by means of oral course, the intestinal epithelium is a big hindrance for the ingestion of hydrophilic drugs, as they can't diffuse across epithelial cells through lipid bilayer cell layers to the bloodstream. In this manner, and the main concern is to improvise the paracellular transport of hydrophilic drugs; thus, a transporter framework is expected to save the protein drugs from the unforgiving condition in the stomach and small digestive system, whenever given orally. To overcome this assortment of intestinal permeation enhancers polymers, for example, chitosan (CS) are used .The insulin stacked nanoparticles covered with mucoadhesive CS has the property of prolonging its living arrangement in the small intestine in our body, invades inside the bodily fluid layer, and subsequently mediate temporarily opening the rigid junctions between epithelial cells while turning out to be unstable and broken separated because of their pH affectability and/ or degradability.[30,35]

5.9.5 NANONEPHROLOGY

As per an investigation, about 900,000 patients worldwide experience the ill effects of end-stage renal illness and need treatment by the process of dialysis or transplantation. Nanonephrology is a part of nanomedicine and nanotechnology that tries to utilize nanomaterials and nanogadgets for the determination, treatment, and management of renal sicknesses. It incorporates the following goals. The investigation of protein structures of kidney at the atomic level. To study the cell forms in kidney cell through nanoimaging approaches, nanomedical medicines that uses nanoparticles, nanorobots, and so forth to overcome various kidney diseases. Advances in nanonephrology will be based on discoveries in the above territories that can give nanoscale data on the cell molecular machinery associated with ordinary kidney forms and in neurotic states.[36,37]

Scientists have built up a human nephron filter (HNF) that would inevitably make conceivable a continuously working, wearable or implantable artificial kidney. The HNF is the primary application in developing a renal substitution treatment (RST) to potentially dispense with the requirement for dialysis or kidney transplantation in last-level renal malady patients. The HNF uses an extraordinary film framework created through applied nanotechnology. In this nanonephrology, the base up nanotechnology is to produce the nanoparticles, which is the get together of new particles or taking atoms and assembling them into new machines. The essential idea of the device is that, it contains two films in an attempt to imitate the ordinary nephron so the way it works is blood first streams over the first membrane, which we state as the G layer and is configured to mirror the capacity of the glomerular basement membrane, so it's a completely permeable membrane removing solute up to the subatomic weight of albumin. The ultrafiltrate that is created by blood passing over this layer at that point ignores a second film we state as the T layer, which is presumably the basic piece of the gadget. This is meant to imitate the cylindrical layer in the renal tubules. The rounded film, the T membrane, has been grown with the goal that it will reabsorb all of those substances we need to hold, some sodium, some potassium, calcium, a smidgen of phosphorous.

5.10 SOFT ROBOTICS

Soft robotics is another part of biorobotics that helps in designing and building robots from uniquely agreeable materials, like the one which are present in human body.[38] Soft robotics gets to know in better ways of how a human body moves and get habitual to the environment variations.

On the other hand, robo gadgets designed by immovable items, and soft robots helps in increasing the flexibility and adaptability for completing any day-to-day work with an ease, and stays cautious while working around human beings.

Soft robotics technology has a few qualities which seem to be very useful and require designing robo-gadgets in surgical biorobotics field. Initially, the robot bodies which are made up of delicate materials can follow its surrounding environment, decreasing capability of harming tissue or organs. Also for, the flexible attributes of delicate robots may give the robotics endoscope a mobility to follow inside patient lumen. And, the delicate objects might have the solidness customizable properties that enable the robotic technology endoscope to solidify planned piece of its body. In conclusion, the vast majority of the delicate materials utilized in soft robotic technology can be used inside MRI bore.[39] This may provide an MR-compatible component for endoscopes which can come over of the endoscope restriction issue.

5.10.1 ADVANTAGED AND APPLICATIONS OF SOFT ROBOTICS IN MEDICAL FIELD

5.10.1.1 FIELD AS AN ASSISTANT AT SURGERIES

Soft robots would be used in the clinical calling, explicitly for intrusive medical method used in treatment. Soft robots could become a helping hand in medical procedures due to their shape evolving characteristics. Body design variation is significant as a soft robot may find and learn about a lot many shapes and structures in the human body by making changes in these shapes and structures. And it could be cultivated using fluidic activation.

5.10.1.2 EXOSUITS

Soft robots are being highly apt for production of adaptable exosuits designing and utilizing, for patients who are recovering, helping the old, or generally for upgrading the client's quality. A group from Harvard made an exosuit utilizing these materials so as to invigorate the benefits of the extra gave by an exosuit, without the disservices that accompany how inflexible materials confine an individual's day-to-day development.

The exosuit are systems that are made up of hard metallic substances fitted with mechanized muscles to duplicate the wearer's quality. Additionally called exoskeletons, the automated suits metal system fairly reflects the wearer's interior skeletal structure. The suit causes lifted items to feel a lot lighter and now and then even weightless, decreasing wounds and developing consistence.

5.10.1.3 COLLABORATIVE ROBOTS

Generally, producing robots was secluded from man laborers due to the safety reasons, as an in-mobile robotic instrument slamming inside human could undoubtedly prompt injury due to rapid motion change with high speed of the robot. Be that as it may, delicate robots may work close by human being without any danger, like a crash about the consistent idea of the robot would forestall or limit any potential injury.

5.10.1.4 BIOMIMICRY

A utilization of biomimicry by means of delicate mechanical autonomy is in sea or space investigation. In the quest for extraterrestrial life, researchers need to find out about extraterrestrial waterways, as it is a wellspring of lives on Earth. Delicate robo-gadgets might be utilized to copy ocean animals that may productively move under water. Such an undertaking was endeavored by a group at Cornell in 2015 under an award through NASA's Innovative Advanced Concepts (NIAC). The group designed a structure for a delicate robot which can copy a lamprey or cuttlefish in the manner it may move to get submerged, so as to productively investigate the sea beneath the ice layer of Jupiter's moon, Europe. Be that as it may, investigating a waterway, particularly one on another planet, accompanies an exceptional arrangement of mechanical and materials challenges.

5.11 CONCLUSION

In this scope of book, we conclude that this is the era of new beginning by the combination of biology with robotics technology designing new devices, gadgets, instruments which helps in diagnosing and treating incurable diseases at low cost and at high speed there are several benefits of bio robotics, nanotechnology, soft robotics in various areas of medical field. And these developments in the field of biomedical signal processing and controlling is giving birth to a whole new world with better future and results for medical field. We studied in detail about the various technologies being introduced for processing and controlling biological signals.

KEYWORDS

- **biorobotics**
- **nanorobotics**
- **nanotechnology**
- **bionics**
- **cybernetics**

REFERENCES

1. Vincent, J. F. V.; Bogatyreva, O. A.; Bogatyrev, N. R; Bowyer, A.; Pahl, A. K. Biomimetic: Its Practice and Theory. *J. R. Soc. Interface*. **2006,** *3*(9), 471–482. DOI: 10.1098/rsif.2006.0127. PMC 1664643. PMID16849244 (22 August 2006).
2. Bhushan, B. Biomimetics: Lessons from Nature-an Overview. *Philos. Trans. Royal Soc. A Math. Phys. Eng. Sci.* **2009,** *367*(1893), 1445–1486. Bibcode:2009RSPTA.367. 1445B. DOI:10.1098/rsta.2009.0011. PMID19324719 (15 March 2009).
3. Knight, T. Idempotent Vector Design for Standard Assembly of Biobricks, 2003. hdl: 1721.1/21168.
4. Shetty, R. P; Knight, T. F.; Endy, D. Engineering Bio Brick Vectors from Bio Brick Parts. *J. Biol. Eng.* **2008,** *2*(5), 5. DOI:10.1186/1754-1611-2-5. PMC 2373286. PMID 18410688 (14 April 2008).
5. Kevin, K. *Out of Control: The New Biology of Machines, Social Systems and the Economic World*. Addison-Wesley: Boston, 1994. ISBN 978-0-201-48340-6. OCLC 221860672, (1994).
6. Mehrali, M.; Bagherifard, S.; Akbari, M.; Thakur, A.; Mirani, B.; Mehrali, M.; Hasany, M.; Orive, G.; Dolatshahi-Pirouz, A. Flexible Bioelectronics: Blending Electronics with the Human Body: A Pathway Toward a Cybernetic Future (Adv. Sci. 10/2018). *Adv. Sci.* *5*(10), 1870059. DOI:10.1002/advs.201870059. ISSN 2198-3844. PMC6193153 (October 2018).
7. Heylighen, F. Foundations and Methodology for an Evolutionary World View: A Review of the Principia Cybernetica Project. *Found. Sci.* **2000,** *5*, 457–490.
8. Korotayev, A. V.; LePoire, D. J., Eds. *The 21st Century Singularity and Global Futures*, 2020.

9. Vaughn, J. R. Over the Horizon: Potential Impact of Emerging Trends in Information and Communication Technology on Disability Policy and Practice. National Council on Disability, Washington DC, 2006; pp 1–55.
10. Ghosh, A.; Fischer, P. Controlled Propulsion of Artificial Magnetic Nanostructured Propellers. *Nano Lett.* **2009,** *9*(6), 2243–2245. DOI: 10.1021/nl900186w. PMID 19413293.
11. Sierra, D. P.; Weir, N. A.; Jones, J. F. Are View of Research in the Field of Nanorobotics. U.S. Department of Energy – Office of Scientific and Technical Information Oak Ridge, TN. SAND2005-6808; pp 1–50. DOI: 10.2172/875622(2005).
12. Kong, L. X.; Peng, Z.; Li, S. D. Nanotechnology and its Role in the Management of Periodontal Diseases. *Periodontol 2000* **2006,** *40*, 184–196.
13. Soto, F.; Chrostowski, R. Frontiers of Medical Micro/Nanorobotics: in vivo Applications and Commercialization Perspectives Toward Clinical, 2018.
14. Feynman, R. P. There's Plenty of Room at the Bottom, Engineering and Science (California Institute of Technology), February 1960; pp 22–36. Reprinted in *Nanotechnology: Research and Perspectives*; Crandall, B. C., Lewis, J., Eds.; MIT Press, 1992; pp 347–363. In *Miniaturization*; Gilbert, D. H., Ed.; Reinhold: New York, 1961; pp 282–296. See also: http://nano.xerox.com/nanotech/feynman.html.
15. Taniguchi, N. Current Status in, and Future Trends of, Ultraprecision Machining and Ultrafine Materials Processing. *Ann. CIRP 32* **1983,** *2*, 573–582. Taniguchi, N., Ed.; Nanotechnology: Integrated Processing Systems for Ultra-Precision and Ultra-Fine Products, Oxford University Press: Cambridge, 1996.
16. Mikhail, V. *Volkenstein, Molecules and Life: An Introduction to Molecular Biology*; Plenum Press: New York, 1970.
17. Burks, A. W.; Neumann von, J., Eds. *Theory of Self-Reproducing Automata*; University of Illinois Press: Urbana IL, 1966.
18. Eric Drexler, K. Molecular Engineering: An Approach to the Development of General Capabilities for Molecular Manipulation. *Proc. Natl. Acad. Sci. USA* **1981 Sep,** 78, 5275–5278.
19. Drexler, E. Mightier Machines from Tiny Atoms May Someday Grow, Smithsonian, 13 November 1982; pp 145–155.
20. Gonzalez, C. More Than Meets the Eye: The Future of Bio-Robotics, Feb 28, 2018.
21. Li, J.; Esteban-Fernández de Ávila, B.; Gao, W.; Zhang, L.; Wang, J. Micro/Nanorobot[14]s for Biomedicine: Delivery, Surgery, Sensing, and Detoxification. *Sci. Robot.* **2017,** *2*, eaam6431.
22. Nelson, B. J.; Kaliakatsos, I. K.; Abbott, J. J. Microrobots for Minimally Invasive Medicine. *Annu. Rev. Biomed. Eng.* **2010,** *12*, 55–85.
23. Freitas, R. A. Jr. Nanodentistry. *J. Am. Dent. Assoc.* **2000,** *131*(11), 1559–1565.
24. Stephen C Bayne. Dental Biomaterials: Where are we and Where are we Going? *J. dent. Educ.* **2005,** *69*(5), 571–585.
25. Roberson, M. T.; Heymann, O. H.; Swift, E. J. Jr. *Biomaterials Sturdevant's Art and Science of Operative Dentistry*; 5th ed.; Mosby Co, 2006; pp 137–139.
26. Gumusderelioglu, M.; Mavis, B.; Karakecli, A.; Kahraman, A. S.; Cakmak, S.; Tigli, S.; Demirtas, T. T.; Aday, S. *The Nanotechnology Concept in Dentistry*, 2007; p 479.

27. Ashammakhi, N.; Ndreu, A.; Yang, Y.; Ylikauppila, H.; Nikkola, L.; Hasirci, V. Tissue Engineering: a New Take-Off Using Nanofiber-Based Scaffolds. *J. Craniofac. Surg.* **2007,** *18*(1), 3–17.
28. Rachel, M. S.; Madalina, C.; Amy, E. R.; David, E. C. A Multiwalled Carbonnanotube/Dihydropyran Composite Film Electrode for Insulin Detection in a Microphysiometer Chamber. *Anal. Chim. Acta.* **2008,** *609*(1), 44–52.
29. Awadhesh, K. A.; Lalit, K.; Deepa, P. K. T. Applications of Nanotechnology in Diabetes. *Digest J. Nanomat. Biostr.* **2008,** *3*(4), 221–225.
30. Smyth, S.; Heron, A. Diabetes and Obesity, the Twin Epidemics. *Nat. Med.* **2006,** *12*, 75–80.
31. Krauland, A. H.; Guggi, D.; Bernkop-Schn¨urch, A. Oral Insulin Delivery, the Potential of Thiolated Chitosan-Insulin Tablets on Non-Diabetic Rats. *J. Control. Release* **2004,** *95*, 547–55.
32. Borchard, G.; Lueßen, H. L.; DeBoer, A. G.; Verhoef, J. C.; Lehr, C. M.; Junginger, H. E. The Potential of Mucoadhesive Polymers in Enhancing Intestinal Peptide Drug Absorption. III, Effects of Chitosan-Glutamate and Carbomer on Epithelial Tight Junctions in vitro. *J. Control. Release* **1996,** *39*, 131–138.
33. Kotz′e, A. F.; Lueßen, H. L.; deLeeuw, B. J.; deBoer, A. G.; Verhoef, J. C.; Junginger, H. E. Comparison of the Effect of Different Chitosan Salts and N-Trimethyl Chitosan Chloride on the Permeability of Intestinal Epithelial Cells(Caco-2). *J. Control. Release* **1998,** *51*, 35–46.
34. Lamprecht, A.; Koenig, P.; Ubrich, N.; Maincent, P.; Neumann, D. Low Molecular Weight Heparin Nanoparticles, Mucoadhesion and Behaviour in Caco-2 cells. *Nanotechnology* **2006,** *17*, 3673–3680.
35. Ramadas, M.; Paul, W.; Dileep, K. J.; Anitha, Y.; Sharma, C. P. Lipoinsulin Encapsulated Alginate-Chitosan Capsules: Intestinal Delivery in Diabetic Rats. *J. Microencapsul.* **2000,** *17*, 405–411.
36. Debjit, B.; Chiranjib, R.; Margret, C.; Jayakar, B. Role of Nanotechnology in Novel Drug Delivery System. *J. Pharm. Sci. Tech.* **2009,** *1*(1), 20–35.
37. Reddy, J. R. K.; Sagar, E. G.; Prathap, S. B. C.; Ramesh, K. B.; Chetty, C. M. S. Nanomedicine and Drug Delivery–Revolution in Health System. *J. Global Trends Pharm. Sci.* **2011,** *2*(1), 21–30.
38. Trivedi, D.; Rahn, C. D.; Kier, W. M.; Walker, I. D. Soft Robotics: Biological Inspiration, State of the Art, and Future Research. *Appl. Bion. Biomech.* **2008,** *5*(3), 99–117.
39. Polygerinos, P.; Correll, N.; Morin, S. A.; Mosadegh, B.; Onal, C. D.; Petersen, K.; Cianchetti, M. ; Tolley, M. T.; Shepherd, R. F. Soft Robotics: Review of Fluid-Driven Intrinsically Soft Devices; Manufacturing, Sensing, Control, and Applications in Human-Robot Interaction. *Adv Eng Mater*. **2017,** *19*(12), 1700016.

CHAPTER 6

Deep Learning-Based Object Recognition and Detection Model

AMAN JATAIN*, KHUSHBOO TRIPATHI, and SHALINI BHASKAR BAJAJ

Department of Computer Science, Amity University, Haryana, India

Corresponding author. E-mail: amanjatainsingh@gmail.com

ABSTRACT

The chapter discusses the design and development of a novel object detection model keeping the motive that anyone and everyone with little to no prior knowledge can easily use it. The object detection is a part of computer vision which deals with detection and localization of objects in images or videos. These systems can be implemented for surveillance, keeping track of goods, robotics, medical field, self-driving cars, and much more. A lot of organizations are using it for various purposes such as it is used by NASA for analysis for deep space analysis, self-driving cars use it for making accurate decisions, and airports use it for security and surveillance. Although it has numerous use cases, but because of complex implementation and need of expertise in the field, it is not being widely used currently. In this chapter, we have made an effort to automate most of the process that a user should do in order to train and use models for object detection. The most complex part of using these technologies is training custom models, and only people having expertise in field are doing it. With the system designed by us, anyone would be able to train custom models on their own and they don't have to do much for it. Normally, first users have to gather large amount of images and then annotate them

Deep Learning in Visual Computing and Signal Processing. Krishna Kant Singh, Vibhav Kumar Sachan, Akansha Singh & Sanjeevikumar Padmanaban (Eds.)

one by one, which is a tedious and time taking process. One has to do this for thousands of images and each of those images can have multiple instances of various objects to train a model to give a good accuracy. We are automating the whole process with the help of AWS recognition module of AWS. You only look once (YOLO) object detection model is used for automation, as it is fast and accurate at the same time which gives it edge over the other models available for use. This model is based on deep learning to learn through the images provided by user in training phase and then later on can be used to perform detection on images or real-time videos. A graphic user interface (GUI) is also developed, so the users can easily interact with the system and train their own custom models and use as they want it to be used. Data are sent over the AWS servers and retrieve the results using a block of python code, then the response collected from AWS is manipulated and the data are extracted that is required to annotate the images and then processing is done according to the need of our target model. This format can be different for various models which can be used for object detection.

6.1 INTRODUCTION

Objects have been an important field since the rise of the artificial intelligence, detection and recognition because of its possible uses in the real-world scenarios. It can be used in driverless cars, surveillance systems, automation of various processes, and also but not limited to security purposes. These systems involve not only recognizing and classifying objects but will also be finding the detected objects along with their locations.[1] Also, the significant increase in processing power of devices in recent years plays an important role in the rising demand of object detection. Methods for object detection can be generally classified into two approaches, that is, machine learning-based approach and deep learning-based approaches. For machine learning-based approaches, we first need to define a feature map using one of the methods below and then using techniques to perform the localized objects in a given frame. On the other hand, the deep learning-based approaches are able to perform end-to-end object detection on their own without the need of specially designing the feature maps as in the case of machine learning-based approach. It uses convolutional neural networks (CNNs). Some of the commonly used methods using both the methods are: (1) Machine learning approaches—Viola-Jones (VJ) object detection framework based on HAAR features, scale invariant feature transformation (SIFT), histogram of oriented gradients (HOG) feature. (2) Deep learning approaches—Region proposals (R-CNN, fast R-CNN, faster R-CNN), you only look once (YOLO), single shot refinement neural network for object detection (Refinenet), single shot detector (SSD).

6.1.1 MACHINE LEARNING-BASED OBJECT DETECTION

This approach can be divided in two phases, the training phase and the detection phase. The steps during the training phase are:

- Collect training data (here images) and organizing it according to their respective class.
- Then preprocess the images and try to separate the foreground and the background of the images.
- After this step feature extractors like SIFT or HOG are used to extract the features and thus simplify the image. It makes the detection faster.
- Then the training data are passed to a classifier to train on the data.
- Once the model is trained, then the testing data is passed to check the accuracy of the newly trained model.

After the training is completed, the newly trained models are saved on devices locally and can be used at some later time. When these saved models are used to process an image or a video feed, then this is known as detection phase. The steps involved in this phase are:

- Rescale the image to the size at which our model is been trained.
- Use sliding window to locate the region of interests (ROI).
- Take those ROI and extract the feature using various feature extractors.
- Pass the feature map to the trained model.

The model will try to classify the object by going through the feature map provided.

6.1.2 DEEP LEARNING-BASED APPROACH

There are various instances discussed in literature where deep learning is used for object detection. Deep learning is a subset of machine learning in artificial intelligence that has network capable of performing unsupervised learning from data that are unstructured or unlabelled. It imitates the working of human brain in processing data and searching for patterns in given data for use in decision-making. These neural networks are built like human brains, with neuron nodes connecting each other and forming a web like network, where output of each layer of nodes is feed as input of next layer of neurons/nodes. While traditional models learn from data in linear way, deep neural networks learn from data in a hierarchical way. The trade-off here is on speed, it is relatively slower than the machine learning approach and also requires a lot more resource to run. But since we are mostly dealing with the real-world scenarios where accuracy is lot more important than speed. Our devices can be improved by investing into them but the accuracy cannot be improved once the system is ready to use and is implemented on a wide scale. It uses convolutional layers to create CNNs and then perform everything, from feature extraction to object classification and localization on their own. There are various layers which are specifically designed for a certain type of task at various layers in these networks. Also each layer processes the input in a different way.

6.1.3 APPLICATIONS OF OBJECT DETECTION

Today, one can perform real-time object detection easily which seemed to be impossible a decade ago. There have been a lot of models which can be used for object detection like YOLO, SSD, R-CNN, etc.[2] One of the most common uses of object detection, which everyone of us must have seen at some point of time is, tracking an object and predicting its path.[3] This is used in various sports for the analysis and sometimes to make correct decisions. Some of the fields where object detection is in much demand nowadays are:

- **Surveillance**

  Anyone with knowledge of deep learning and computer vision can easily automate the surveillance systems. These save manpower and chances of error because of human failure. It is still a field of research and it is growing at a rapid rate because of increased number of security cameras all over the world, and there can't be enough manpower to monitor each one of them manually.

- **Robotics**
  Object detection is being very widely used in robotics. With rise of humanoid and AI powered robots, there is a need of technology through which the robots can understand their surroundings for a seamless working. They need to identify which of the objects they should avoid and which ones are actually useful for them. This is where object detection comes into play.
- **Self-driving cars**
  Self-driving cars needs something to perceive its surroundings and process the video feeds coming from the various onboard cameras on a car to make suitable detection in real time. These cars use object detection models to do this.
- **Space research**
  NASA is building an object detection model which will search for exoplanets in the picture captured by telescopes. It will greatly improve the accuracy of the process and will also reduce the time taken to do the same by a human.
- **Medical field**
  Medical images can be very complex and require very highly qualified and experienced doctors for evaluation. Still there are chances of human error while inspecting these images. So, these days, object detection modes like U-Net is used for processing such images.

Even though object detection models are available for free in public domain, it can be complicated to get started with these models as they need either a high system configuration for acceptable performance or high technical knowledge is a requisite. Therefore, in the current scenario when there is a need to train the models from the scratch, we need to build annotation to manage the data in a proper way.[16] These annotations are used to determine the location and class of objects in an image. There are five columns in these annotation files, top left x and y coordinates, width, height of object, and the class to which it belongs (generally an integer value). This class value is further mapped to a particular class name which is set accordingly by user. All these data are to be created manually by user again and again and give the class name to each object which is going to be used for the training purpose. The main objective of this research is to automate the whole process in which user doesn't have to mark down each of the objects in each image. In other words, there is no need to make the annotations manually and the computer would do it on its own. Also during the training a model, any number of classes can be selected from the saved data and a model will be trained for that data and that will be stored for later use to can select any of the saved model from a dropdown list and use it to apply on the current video feed.

6.2 RELATED WORK

In literature, object detection methodology is mainly discussed relating to two techniques, that is, machine learning and deep learning-based object detection methods while other being the deep learning-based object detection methods.

6.2.1 MACHINE LEARNING DETECTION METHODS

VJ: VJ object detector was the first real-time detection of human faces. Named after P. Viola and M. Jones,[1] this technique was propounded in the year 2001. The VJ detectors ran on a 700 MHz Pentium III CPU that worked on a gigantic faster rate than any other algorithm prevailing in that period. The detectors follow the scheme of detection, that is, sliding windows: to go through all possible locations and scales in an image to see if any window contains a human face. The characteristics of the VJ detectors that make it reliable algorithm are: (1) Robust, (2) Real time, (3) Face detection. However, it seemed to be a very easy process of just detection of human faces but the calculations on which it was operating were tedious for the computer in those days.[2] The working of the detectors had been improved with integrating the three techniques: (1) Integral image: It speeds up the box filtering process; the integral image makes the computational complexity of each window in VJ detector independent of its window size. (2) Feature selection: Adaboost algorithm was being used by the developers to select a small set of features that are mostly helpful for face detection. This algorithm constructs a "strong" classifier as a linear combination of weighted simple "weak" classifiers. (3) Detection cascades: This was introduced to reduce computational overhead by spending less computations on background windows but more on face targets. In cascading, each stage consists of a strong classifier. So, all the features are grouped into several stages where each stage has certain number of features.

HOG: HOG[3] is one of the feature descriptor used to detect objects. This concept gained the worldwide attention in 2005 when N. Dalal and B. Triggs presented it at the Conference on Computer Vision and Pattern Recognition (CVPR). The foremost purpose of the HOG detector was to focus on pedestrian detection in static images. The HOG detector adjusts the size of the input images which come in the contrasting sizes. On the other hand, the size of the detection window remained uninterrupted. The distribution of intensity gradients or edge directions describes local object appearance and shape within an image. The image is divided into small connected regions called cells. The histogram of gradient compiled directions of the pixels within each cell. The list of algorithms implemented in HOG detector is: gradient computation, orientation binning, descriptor blocks, block normalization, object recognition.

DPM: DPM was upgradation of the HOG by P. Felzenszwalb[4] in the year 2008 which was accompanied by a number of varieties by R. Girshick.[5–8] DPM detector has a strategy of divide and conquer in which at training phase a weakly supervised learning of befitting way to break down an object occurs and later the conjecture is grouping the different object parts for detection. For example, the detection of a "car" can be done by determining its window, its wheels, and shape. This is known as "star-model". Later R. Girshick further improved star-model to the "mixture-model" so that real-world object can be detected with more variations. To improve detection accuracy, some major techniques were formulated such as "hard negative mining," "boundary box regression," and "context priming".

6.2.2 DEEP LEARNING DETECTION METHODS

RCNN: Ross Girshick et al.[9] proposed a method of object detection in 2014. The method was easy; firstly, detect region proposal[10] using selective search then each object candidate box is arranged to a fixed size image and these fixed size images are the input to the CNN model trained on ImageNet to extract the features. At last using linear SVM classifier presence and category of an object are predicted. RCNN remarkably boosts the performance on VOC07, with a large improvement of mean

Average Precision (mAP) with 33.7% (DPM-v5[11]) to 58.8%. RCNN has some disadvantages also, such as classifying 2000 region proposals per image required huge amount of time to train the network. Real-time implementation is not possible as it required around 47 s for each test image.

SPPNet: Kaiming et al.[12] proposed Spatial Pyramid Pooling in deep convolutional network. A Spatial Pyramid Pooling layer was introduced which eliminate the fixed size input image requirement of RCNN. In SPPNet, fixed length representation is generated irrespective of image size/scale. SPPNet also shows a good impact on object detection. It determined the feature maps at once from the entire image and then pool features in sub images use for generating fixed-length representations for training the detectors. This reduced convolutional features computation. This method is 24–102x faster than the RCNN method and achieved a comparable accuracy on Pascal VOC 2005. It also has some drawbacks such as it only focuses on fully connected layers while ignoring other layers and it also has multistage training.

Fast RCNN: Ross Girshick[13] proposed an improvement version of RCNN and SPPNets and named Fast RCNN. A Fast RCNN takes an entire image and a set of object proposals as input. The entire image is processed with the several convolutional network and max pooling layer which produces a convolutional feature map. The ROI pooling layer processes each object proposal and extracts a fixed length feature vector from the feature map. The RoI is a special case of spatial pyramid layer of SPPNets. A sequence of fully connected layers is branched into two sibling output layers. One of these sibling output layers produces softmax probability over K object classes and another layer gives four real-valued numbers for each of the K object classes. Fast RCNN has an ability to train all network weights with back propagation. Fast RCNN improved the mAP from 58.5% (RCNN) to 70.0% and also increased detection speed over 200 times faster than RCNN on VOC07 dataset.

Faster RCNN: Shaoqing Ren et al.[14] introduced an object detection system called Faster R-CNN which has two modules. The first module is a deep fully convolutional network that proposes rectangular regions, and the second module is the Fast R-CNN detector.[13] The main contribution of Faster RCNN is it introduced regional proposal network (RPN). An image of any size is input for Faster RCNN and a set rectangular object proposals, each with objectness scores which measures membership to a set of object classes vs background. Faster RCNN makes a unified, deep learning-based object detection system that runs nearly at real-time frame rates. In Faster R-CNN, both region proposal generation and objection detection tasks are all done by the same conv networks. With such design, object detection is much faster.

You Only Look Once (YOLO): YOLO is a deep learning-based neural network used for performing object detection in real time. The algorithm applies a single neural network to the full image, which is then divided into regions and then it predicts the objects in different regions as well as their probabilities. It consists of 75 convolutional layers and 31 other layers. So in total there are 106 layers in the network.[11] In the training phase of the network, the images are passed through all the layers one by one, and accordingly filers are created.

6.3 RESEARCH METHODOLOGY

This research focuses to automate the most of the process for users, who don't have in-depth knowledge of things, but are still interested to use object detection, as it can be used in almost any field. Especially when they have to train their own models as it can be a complex process for some users. Each instance of the object needs to be label, which can be a very time taking and tedious task. It is one of the most important reason why object detection is not very widely used these days. Here, AWS recognition module is used to do labeling. Then the data retrieved from AWS servers is converted to YOLOv3 format so that it can be loaded and compiled when training the model.

YOLO object detection model is selected for implementation of the proposed methodology because of its improved accuracy as compared to its counterpart. YOLO trains on full images and directly optimizes detection performance.[13]

6.3.1 YOLO MODEL TRAINING

To train a YOLO model, data are collected and stored in proper structure, so that the data can be accessed and feed to the model properly when it is being trained. Then annotations are also required to be created for each image. Annotation files contain the location of objects and its corresponding class in image. Each image has its corresponding annotation files in a folder named labels. There are various tools which are available to annotate images according to the various object detection models.

6.3.1.1 ANNOTATING IMAGES ACCORDING TO YOLO FORMAT

To annotate image, one file for each image is required and the file extension will be .txt, the file name will be the same and stored in a separate folder named labels. The format of data stored in these files is:

Class x y width height

There can be more than one row, and the number of rows depends upon the number of objects in the given image. Each of the object have its corresponding class, its top left x and y coordinate, and its width and height stored in these files, through which the model get to know where the object is located in image and try to learn from it. The x and y coordinates, width and height, all are scaled down in a range of 0–1. Some examples of annotated data are:

1 0.716797 0.395833 0.216406 0.147222

0 0.687109 0.379167 0.255469 0.158333

1 0.420312 0.395833 0.140625 0.166667

There are various open source software's available online through which images can be annotated in YOLO format, for example, LabelImg, BBOX, etc.

6.3.1.2 STORING DATA IN DIRECTORIES

All the images and all their corresponding labels are collected and then stored in two separate folders named images and labels.

6.3.1.3 CONFIG FILES

This file helps the model to know where to look for some essential data which are required during the testing and detection phase of models. The file contains five fields which are class, train, test, names, and backup.

- The classes field contains the total number of classes for which the model is been trained.
- Train field contains all of the image names which are intended to be used for training purpose.
- Val field stores name of all the images which are intended to use for validation purpose.
- The classes are in form of integers and so to map them to a name we create a file having all of the names of the classes in ordered manner. The names field stores the path to that file.

6.3.1.4 MODIFYING PART OF CODES AND TRAIN THE MODEL

Some values such as the classes are still needed to be updated at the lower part of cfg file and some other lines of codes are needed to be adjusted according to the new number of classes in some other places which is used when the model is compiled and trained. After making all the other adjustments, new custom model is trained using TensorFlow or Keras to use the inbuilt functions to do it or user can write own functions to do the same. To ease and make it efficient and faster, Keras is used to compile the models as it provides a lot functions and utilities to work with CNNs. There are simple functions for most of the operations, and one can also generate new models by stacking one convolutional layer above another using Keras.

6.3.2 AUTOMATING THE ANNOTATION PROCESS

Generally, when user needs to train a custom YOLO model, a large number of images are annotated which are used by the models while it is learning. These numbers can easily reach to tens or even hundreds of thousand if there are 50 classes or so. Doing all this manually can be a very time taking and hectic work. Normal people generally never bother to do all this on their own, and this is one of the biggest reason why object detection is not being widely used in daily life. We are trying to eliminate this annotation process from the training phase. The key to automate the annotation process is first to detect and classify the objects in each of the images. For this, two strategies are worked upon:

Using image segmentation

- First, the foreground and the background of image are separated.
- Then image segmentation is applied on the foreground of the image, which will give the location of the objects.
- To group all of the similar objects, a classifier is used, and then the annotation files are created accordingly.
- The problem of this method is that it is computationally expensive and very hard to implement.
- In some cases this method can be inaccurate too.

Using pretrained models for a large number of classes

- In this method, pretrained object detection model is used to detect and classify the objects in image.
- Once objects are detected, the coordinates and classes of objects in each image are stored.
- After storing the coordinates, it is scaled down in range of 0–1, which is the proper format for annotation files for YOLO.
- After this, a class map is created, where an integer value is assigned to each of the class and then stored accordingly.
- It is much simpler to implement, but it is difficult to find pretrained models for a large number of class.

6.3.2.1 ALGORITHM

For this, a pretrained model is used, or more like a cloud-based service which can do this for user. AWS Rekognition API's are used that provide multiple features such as labels, custom labels, text detection, content moderation, face detection analysis, patching, and celebrity recognition. We are using the object and scene detection feature of AWS Rekognition. It is trained for a very large number of classes and we can send images to it through the internet and then retrieve the result using python. The **boto3** library of python is specifically designed to work with AWS, and the same is being used to send data and retrieve the results from AWS server.

Algorithm for automating the annotation process using the object and scene detection feature of AWS Rekognition are:

- Setup an IAM user on AWS server and give it administrative access, so that it can utilize all of the services remotely.

- Save the aws_access_key_id and aws_secret_access_key which is used to authenticate with AWS server.
- Establish a connection between the code and the AWS server using the keys saved earlier.
- Use the detect_labels() function of boto3 library to pass the image and store the result in variable.
- The response provided by AWS contains a lot of data which we don't need, so we process the data, and separate the names and coordinates of objects in a separate variable.
- Scale the coordinates in the range of 0–1.
- Now, we replace the class names with integer value and create a word map accordingly.
- Now, user can decide which classes he is going to train for and save only that discarding the rest of data.

6.3.3 YOLOv3 DETECTION

YOLOv3 is better, not faster, stronger than the previous versions of YOLO. Fot it's time, YOLO9000 was one of the fastest and most accurate algorithms, but with continuous improvements of other models such as RetinaNet and SSD, they were outperforming it. Even though it was still the fastest, it still needed to be improved to be still in competition. This was why YOLOv3 was introduced. The speed of YOLOv3 is slightly slower than older version but that speed has been trade-off for the increased accuracy. The older model used to run at 45 FPS on a NVIDIA Titan X, while it runs at only about 30 FPS.

6.3.3.1 DARKNET-53

Darknet is an open source neural network framework written in C and CUDA. It is fast, easy to uninstall, and supports CPU and GPU computation according to the availability. Using GPU for computation, you can greatly increase the computation speed. YOLOv3 uses Darknet-53, which originally has 53 layers and is trained on ImageNet. For the detection phase, 53 more layers are stacked onto the architecture, making the total count of layers to 104. YOLO9000 only had 30 layers in total, and the higher number of layers lead to the YOLOv3 being slower.

6.3.3.2 DETECTION AT VARIOUS SCALES

The newer architecture of darknet-53 uses residual skip connections and upsmapling. The residual skip connections allow gradient to flow through a network directly, without passing through the nonlinear activation functions. The residual connects a virtual bus through which gradients can flow in both directions and thus, can bypass the activation functions along the way. The process of increasing sampling rate of digital signal is known as upsampling. The upsampling layers have no weights associated with it and it doubles the dimension of input. This is the method how we perform the detection at various scales. This helps to detect smaller objects which cannot be detected by the earlier versions. When using YOLOv3, detection takes place at three layers in network each having a different resolution to work with. The eventual output is computed by applying a 1×1 kernel on a feature map. The shape of the detection kernel is $\mathbf{1 \times 1 \times (B \times (5 + C))}$. Here, B is the number of bounding boxes on a cell that can be predicted using the feature map, 5 is used here for the four attributes of the bounding box and one object confidence, that presents how confident the model is that there is a object in that given cell, and C is the number of classes for which the given model is trained. The model performs detection at three scales, the 13×13 layer is responsible for detecting the larger images, the 26×26 layer is used to detect medium sized objects, and the third one, that is 52×52 layer is used to detect very small objects in a given image.

6.3.3.3 SELECTING ANCHOR BOXES

YOLOv3 uses three anchor boxes at each scale. So in total, there are nine anchor boxes because of the three different scales used for detection. These anchor boxes helps us to pinpoint the accurate location of objects in the given image. When we are training our own custom YOLO model, we should use K-means clustering to generate these nine anchor boxes, which are then arranged in the decreasing order of their dimension.

6.3.3.4 HIGHER NUMBER OF BOUNDING BOXES

As discussed above, the YOLOv3 is slower than Yolov2. This is because in YOLOv2 its native resolution is (416×416), the model would predict

$13 \times 13 \times 5$, that is, 845 bounding boxes in total, while in case of YOLOv3 it can detect 10,647 at the same resolution. This higher number of calculation is the result of using nine anchors, three at each scale.

6.3.3.5 LOSS FUNCTION

In the previous versions, the errors are used to be calculated in the form of squared errors. But now the method of calculating the loss is changed and it uses cross entropy error terms. In simpler words, the object confidence and class prediction in YOLOv3 are determined with the help of logistic regression. While training the model, we assign a bounding box for each ground truth box, whose anchor has the maximum overlap with the ground truth boxes. The other of the anchor is suppressed and only the selected one would be the output for given instance.

6.3.3.6 PREDICTION OF MULTIPLE CLASSES

In earlier models, a given instance of an object can only by classified into a single class. The model would consider the given instance of object's class to be the one which have maximum probability. The rest of the classes are suppressed. These cases arise mostly because the condition of mutual exclusion. For example, the same object may be a vehicle and a car, to remove this problem, each class score is predicted using logistic regression and it can be classified to any number of classes if it crosses the threshold value for the same instance.

6.3.4 ARCHITECTURE OF YOLO

The model consists of 105 CNN layers. The network can be divided into two parts; the first one is used to process the image in such a way that it decreases the extra data while maintaining the useful data. The latter half of the object is used to detect and classify the objects. The later half can further be divided into three parts, and in each of these parts the detection is performed at three different scales to cover and handle the change with a different size of object.

There are multiple possible ways in which these models can be loaded for compilation and then it can be transferred to other machines for various purposes. One of those methods is storing it in the form of cfg files. Here, each of the layers are specifically defined by some of their basic parameters, which can be later on compiled into various machines and programming languages.

- Here, the [Convolutional] represents that it starts a new convolutional layer.
- The second parameter is batch normalization, which helps us to smoothen the training process and make it faster. It sends data to the layers in the form of a batch instead of sending all of it one by one.
- The next parameter is a filter which is a set of adjustable weights which learns based on backpropagation methods. Each filter is storing a single trait or pattern from the images which can be later on compared to detect similar type of patterns. These filters slide from left to right by coming down one row after reaching the end of the current row. This happens until it slides over the whole image and extracts all the data that can be extracted.
- The fourth parameter is the size of filters that is used to iterate over the image.
- The fifth parameter, stride is the number of pixels the filters shift after each time over the given image matrix. A stride of 1 means it shifts only one pixel each time, but stride of 2 means that it will skip one pixel and move to the second pixel each time.
- The pad parameter is used to add an extra row or column at each edge. This is done so that after applying the filters, the resolution of image remains the same. Otherwise if we are using a filter of size n, then after applying the filter at a layer, the resolution of the input image would decrease by n-1. It's not necessary but it is used in some levels to increase the efficiency of the model.
- The last parameter is activation which stores the type of activation function that will be used for the corresponding layer. There are various types of activation functions available and ready to use and we can select one according to our need.

The last layer of the YOLO is a bit different than the others and this contains some other parameters such as anchor boxes which help to locate the objects, classes that store the number of classes the model is trained for, threshold value above which the detected object would be considered for the final output and some other parameters. User also needs to change these values according to the need if training a custom model.

6.3.5 GUI

A graphic user interface (GUI) is also build so that users can deal with all of the functionalities easily. It also helps the users who don't have any prior knowledge in the domain. The GUI would be designed using TKinter library of python and there are total three types of options provided to the user on the home menu.

- **Add Data**
  Using this, user can add one or more new object classes to preexisting data. The two approaches on which this would be implemented is discussed in the next section.
- **Train Model**
  The user can train a model by selecting any number of categories. This can be done by fetching all of the classes for which data are available and populating them in the form of checkboxes. Just mark all of the checkboxes you need to train for and some of the basic parameters can also be set to train a model to customize it according to ourselves.
- **Perform Object Detection**
  A new window will open and there would be a dropdown list populated with all of the existing models. User can select any of the models and perform the object detection using it.

6.4 RESULT AND DISCUSSION

With the proposed system, one can train their own custom object detection models and use them as without having a lot of experience in its respective field. The created system provides users the capability to train their own custom models by providing a video file related to the classes for which the user wants to train the model. The system would then generate the dataset from compiling the frames from videos and then automatically annotate those images in newly created dataset. User can select number of images processed from the given video which will be then sent to AWS servers and passed on to AWS Rekogntion service, which is a computer vision-based service hosted by AWS. There, the processing of data is performed, and the response is directed back to us. Then, we use Pandas data frames to store data in proper structure.

Since there can be instances of other class in image, we once again confirm the classes from the user on which he wants to train the custom model. After taking the input, the data of the classes are removed, which is not required by user and then a class map to map each of the class to a particular integer value is generated, which is used for training and detection later on. Then the name of the classes are replaced by their corresponding integer values in the data frame and at the same time, the other values which are top left X, top left Y, width and height are also rounded of up to six digits. All of these four values lie between 0 and 1 where 0 represents the starting of the image on x or y axis while 1 represents the extreme end. Now, the annotation files for each image is created with the same name but extension as .txt in a folder named labels in the same directory.

After completing all the steps discussed, then this newly created dataset is used for a new model. Usually images are needed in range of thousands to tens of thousands which can be a very pain stacking and lengthy process so by automating the process the amount of work done by the user is reduced and at the same time also the speed of the detection is increased by providing users to train a new model on a limited number of classes. The higher the number of classes a model is trained for, it will need more number of calculations and operations. In general, the models available are trained on 80 classes, that is, $\mathbf{(3 \times (5 + 80))}$ which results as 255 filters in total. If the models are trained for only 4 of the classes then the number of filters will only be 27 which are a lot lower than 255 filters. For testing purpose, a dataset is created with only three of the images. The user selected four of the classes to train the model, show, car, person, and bus.

Here, all the images are sent to the server, and from retrieved result the detected objects are, for example, shoe, suit, person, jeans, jacket, wheel, car, bus. From all of the classes selected by user, four of them are: its target classes that is person, car, bus, shoe. Then all of the other instances of objects except those four are removed from the data frame and these are stored in drop indexes line. The number of instances of corresponding class can also be checked in a given image. In the end, the class map is generated on the home folder for the corresponding dataset. Once the model is trained and saved, the other parts of the system can be used to perform real-time detection.

6.5 CONCLUSION AND FUTURE WORK

In today's era, it is difficult to imagine even a mediocre smartphone without abilities as object detection and for a photo cloud storage features as object detection is a crucial aspect. From taking baby steps in 1970–1980s to exponential growth since 2000s, the field has seen miraculous advancement. It is not uncommon to see some of these technologies entering consumer market. All of us have seen the potential of such technologies as in Tesla's self-driving car; although, we are still far from achieving human-level performance, particularly in the wild world. We are carefully placing object detection technology in the ever growing number of places; however every possibility suggests that it is about to explode. As mobile robots, and in matter-of-course autonomous machines, are starting to extensively forming the front, the prerequisiteness of object detection systems is acquisitioning more preponderancy. Such technologies will also open other frontiers as nano-bots armed with object detection that will delineate the uncharted territory as ensepulchered depths of the unsounded ethereal apropos Davy Jones's locker. In all such cases, it only makes researcher more curious because it only means one thing that there is so much more to be discovered. Object detection is done in Hollywood movies and all indications suggest that it will continue to do so. There are various fields where object detection might be a game changer, for example, creating cite guides, powering self-driving cars, boosting augmented reality applications, and gaming. Organizing one's visual memory, improving iris recognition, cattle counting, detecting, tracking and buildings, vehicle for urban planning, car park management, containers, and ship tracking that supports managing harbors, logistics. Theses daring applications of computer vision that might belong to a science fiction novel—but are getting very close to reality today.

KEYWORDS

- **artificial intelligence**
- **computer vision**
- **convolution neural network**
- **deep learning**
- **machine learning**
- **object detection**
- **you only look once (YOLO)**

REFERENCES

1. Viola, P.; Jones, M. In *Rapid Object Detection Using a Boosted Cascade of Simple Features*. Proceedings of International Conference on Computer Vision and Pattern Recognition, 2001; pp I–I.
2. Chandrappa, D. N.; Akshay, G.; Ravishnakar, M. Face Detection Using a Boosted Cascade of Feature Using Open CV. *Wirel. Netw. Comput. Intell.* **2012,** *292*, 399–404.
3. Dalal, N.; Triggs, B. In *Histogram of Oriented Gradients for Human Detection*, IEEE Computer Society Conference on Computer Vision and Pattern Recognition, San Diego, 2005; pp 886–893.
4. Felzenszwalb, P.; McAllester, D.; Ramanan, D. In *a Discriminatively Trained, Multi-scale, Deformable Part Model*, International Conference on Intelligent Computation Technology and Automation, IEEE, 2008; pp 8–15.
5. Felzenszwalb, P. F.; Girshick, R. B.; McAllester, D.; Ramanan, D. Object Detection with Discriminatively Trained Part Based Models. *IEEE Trans. Pattern Anal. Mach. Intell.* **2010,** *32*(9), 1627–1645.
6. Felzenszwalb, P. F.; Girshick, R. B.; McAllester, D. In *Cascade Object Detection with Deformable Part Models*. Proceedings of International Computer Society Conference on Computer Vision and Pattern Recognition, IEEE, San Francisco, 2010; pp 2241–2248.
7. Girshick, R. B. From Rigid Templates to Grammars: Object Detection with Structured Models, Ph.D Thesis, University of Chicago, Division of the Physical Sciences, Department of Computer Science, 2012.
8. Girshick, R. B.; Felzenszwalb, P. F.; McAllester, D. In *Object Detection with Grammar Models,* Proceedings of International Conference on Neural Information Processing Systems, ACM, 2011; pp 442–450.
9. Girshick, R.; Donahue, J.; Darrell, T.; Malik, J. In *Rich Feature Hierarchies for Accurate Object Detection and Semantic Segmentation*, Proceedings of International Conference on Computer Vision and Pattern Recognition, IEEE, 2014; pp 580–587.

10. Zhao, Z. Q.; Zheng, P.; Xu, S. T.; Wu, X. Object Detection with Deep Learning: A Review. *IEEE Trans. Neural Netw. Learn. Syst.* **2019,** *30*(11), 3212–3232.
11. Du, J. Understanding of Object Detection Based on CNN Family and YOLO. *J. Phys. Conf. Ser.* **2018,** *1004*(1), 23–25.
12. He, K.; Ren, S.; Zhang, X.; Sun, J. In *Spatial Pyramid Pooling in Deep Convolutional Networks for Visual Recognition*, Proceedings of International European Conference on Computer Vision, Springer, 2014; pp 346–361.
13. Cai, W.; Li, J.; Xie, Z.; Zhao, T.; Lu, K. In *Street Object Detection Based on Faster R-CNN*, Proceedings of 37th Chinese Control Conference, China, 2018; pp 9500–9503.
14. Ren, S.; Girshick, R.; He, K.; Sun, J. Faster R-CNN: Towards Real-Time Object Detection with Region Proposal Networks. *IEEE Trans. Pattern Anal. Mach. Intell.* **2017,** *39*(6), 1137–1149.
15. Vinoth Kumar, B.; Abirami, S., Bharathi Lakshmi, R. J.; Lohitha, R.; Udhaya, R. B. Detection and Content Retrieval of Object in an Image Using YOLO. *IOP Conf. Series Mater. Sci. Eng.* **2019,** *590*, 1–8.
16. Sumit, S. S.; Watada, J.; Roy, A.; Rambol, DRA. In *Object Detection Deep Learning Methods, YOLO Shows Supremum to Amsk R-CNN*, Proceedings of 1st International conference on Computing, Information Science and Engineering, 2020; pp 1–8.

CHAPTER 7

Deep Learning: A Pathway for Automated Brain Tumor Segmentation in MRI Images

ROOHI SILLE*, PIYUSH CHAUHAN, and DURGANSH SHARMA

School of Computer Science, University of Petroleum & Energy Studies, Dehradun, India

**Corresponding author. E-mail: rsille@ddn.upes.ac.in*

ABSTRACT

Deep learning techniques are recently exploited for segmentation of medical images. These deep learning techniques have accomplished state-of-art conduct for automatic medical image segmentation. Image segmentation helps in quantitative and qualitative analysis of the medical images, which leads to diagnosis of various diseases. Manual segmentation of medical images is a laborious task that prevents early diagnosis of diseases. For these particular reasons, automated techniques play a major role in medical image segmentation. The recent research is focused on deep learning algorithms for efficient automatic medical image segmentation. Deep learning algorithms are classified as supervised as well as unsupervised learning. Supervised learning consists of convolutional neural networks (CNNs) and unsupervised learning consists of stacked auto-encoder, restricted Boltzmann's machine (RBM), and deep belief networks. CNNs comprise convolutional layer, pooling layer, dropout layer, and fully connected layer. The convolutional layer carries out the

Deep Learning in Visual Computing and Signal Processing. Krishna Kant Singh, Vibhav Kumar Sachan, Akansha Singh & Sanjeevikumar Padmanaban (Eds.)

convolution operation with a set of kernels along with weights and added biases individually creating a new feature map on the input image at each layer.

Stacked auto-encoder models are designed by insertion of different layers termed to be auto-encoder layers in the form of stack. These layers take image as an input and extracts different features in the form of feature maps in an unsupervised mode lacking labeled data. It is a model that takes input data, gathers feature representations from this, and then uses these feature representations to restructure output data. According to the literature survey, auto-encoder layers are trained independently after which the full network is used to make a prediction by fine-tuning it using supervised training. The neurons present in deep belief nets are densely connected which helps in rapid and accurate learning of a good set of parameters. RBMs are a type of Markov random fields, consisting of input layer or visible layer and a hidden layer that brings hidden feature representation. There are bidirectional connections between the nodes, so latent feature representation extracted from an input vector and vice versa.

This chapter provides an outline on the state of deep learning algorithms for medical image segmentation, highlighting those facets that are frequently useful for brain tumor segmentation. In addition, comparative analysis of the deep learning algorithms is discussed. This concludes that with the different algorithms for segmentation tumor regions from brain MRI images, deep learning has proven to be the most effective in the recent trends.

7.1 INTRODUCTION

Computer-assisted analysis of medical images plays a crucial role at various stages of diagnosis such as classification, registration, detection, segmentation, image enhancement, and image reconstruction. Medical image segmentation helps in detailed diagnosis of various diseases. Due to the complexity of medical images, manual segmentation of medical images is a laborious task. This led to the automated segmentation techniques of medical images. Brain tumor segmentation is crucial for improving treatment possibilities. Brain tumor or lesions are separated from other components of brain images similar to gray matter (GM), white matter (WM), and cerebrospinal fluid (CSF). Deep learning techniques are the recent techniques that accelerated the use of computer-assisted medical image analysis. These techniques refer to the deeply connected neurons with various functions such as convolution and parametric function. These techniques are efficiently used for medical image segmentation. According to the literature survey, deep learning techniques are most competent for segmentation of brain MRI. Deep learning techniques are excelling in extracting the different feature hierarchies.[1] These techniques are categorized into supervised and unsupervised techniques. Supervised learning consists of convolutional neural networks (CNNs) and unsupervised learning consists of stacked auto-encoder, deep belief networks (DBNs), and restricted Boltzmann's machine (RBM).

This chapter focuses on different facets of deep learning algorithms along with its challenges to apply on the medical image analysis. This chapter particularly focuses on deep learning models excluding outdated feature learning methodologies that are applied to the medical image segmentation. The literature survey comprises the various deep learning algorithms utilized for brain tumor segmentation with their performance parameters such as sensitivity, specificity, and dice similarity score.

7.2 BRAIN MRI

Medical images are acquired from different imaging modalities such as X-rays, computed topography scan, and MRI. There are certain advantages and disadvantages associated with the mentioned medical imaging techniques. MRI has following advantages such as high resolution, high signal-to-noise ratio, and capability for soft-tissue imaging[2] as compared to other imaging modalities. CT scans have inferior soft tissue contrast as compared to MRI. For these reasons, MRI is preferred over to analyze and study the brain anatomy and physiology. For analyzing the abnormalities in brain images, brain MRI is segmented into different objects such as GM, WM, CSF, and other lesions/tumors. Brain MRI images comprise different modalities such as spin-lattice relaxation (T1-weighted), spin-spin relaxation (T2-weighted), and fluid attenuation intention recovery (FLAIR). The contrast among these modalities gives nearly a distinctive sign to all tissues separately.[3] Each structure provides altered biological data about brain tissues. Accurate tumor detection is assured only by using all these structures.[4]

While automating the segmentation of brain MRI, three main complications that exist are image noise, fractional volume averaging, and intensity inhomogeneity. Where image noise can alter the intensity of the image which results in false outputs, image intensity inhomogeneity deals with intensity level variations are dependent on the single tissue class variations over the image and images are subjected to partial volume averaging in which images have finite pixel size, that's why they are subjected to this. The pixel intensity of scanned images may be inconsistent with any one class as nomenclature of pixel volumes is a combination of classes in a tissue understudy.

7.3 BRAIN MRI SEGMENTATION ALGORITHMS

Segmentation is a method of splitting image into various sections with related features such as color, texture, boundaries, brightness, gray level, and contrast.[5] The medical image segmentation is a difficult task due to the inconsistent nature of the several sensing modalities. Apart from this, there are so many artifacts such as motion artifacts, which affects the imaging segmentation. In brain images, various intensity artifacts, such as partial volume effect, are existing both in healthy as well tumorous tissues which lead to incorrect segmentation of healthy and unhealthy tissues.[4] The medical images are also effected due to the noise of device and related electronics. This process plays a vital part in mining information from an image for further diagnosis. In brain MRI segmentation, precise and accurate segregation of tumorous region is essential. To enhance the efficiency and accuracy of brain MRI segmentation algorithms, the necessary process as shown in Figure 7.1 followed is:

- Preprocessing
- Skull stripping
- Feature extraction
- Segmentation

7.3.1 PREPROCESSING

Preprocessing is primary step toward precise and accurate segmentation of medical images. Due to the complex nature of the medical images, preprocessing is required to enhance the quality of image. Preprocessing includes contrast enhancement, bias field correction, leveling the inner part, removing the noise, and conserving the edges. There are various methods proposed for preprocessing of brain MRI such as adaptive contrast enhancement using modified sigmoid function, morphological-based contrast-enhancing techniques, and hierarchical correlation histogram analysis.

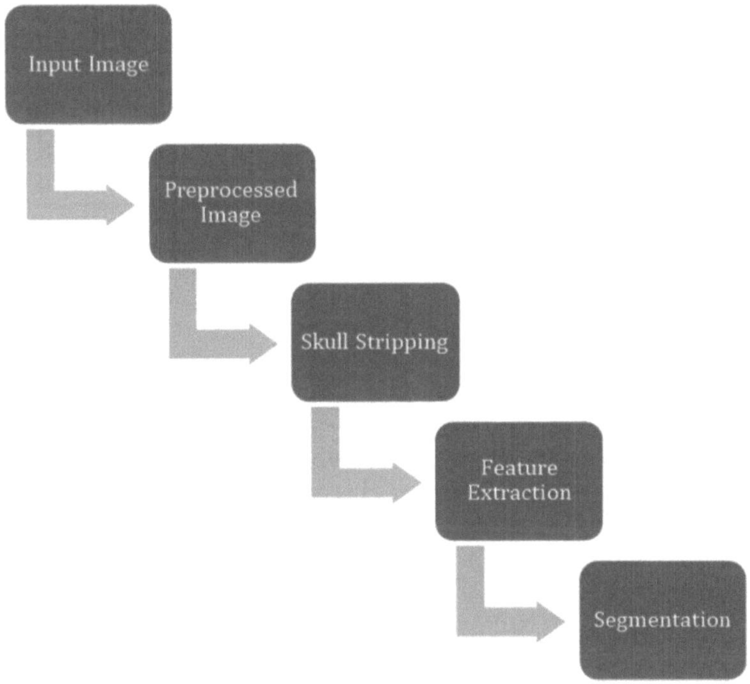

FIGURE 7.1 Segmentation process of brain MRI.

Benson proposed mathematical morphology-based contrast enhancement technique to extract various image components such as shape, edges, and region of interests. Erosion and dilation are the two fundamental operations performed in this technique.[6]

Chen et al. proposed a hierarchal correlation histogram analysis of degree-wise gray scale distribution of pixel intensity for improving the contrast of the image. This method proved to give accurate results based on two evaluation parameters such as average gradient values and peak signal-to-noise ratio (PSNR).[7]

A method based on the average intensity replacement with the help of adaptive histogram equalization is focused upon image enhancement of FLAIR images, performed by adjusting the intensity and contrast. The preprocessed images are evaluated based on PSNR, average gradient, and mean squared error (MSE), which proved this algorithm better than the existing methods. The accuracy of results can be increased by the edge information affecting the WMH regions.[8]

The accuracy of contrast enhancement algorithms is evaluated on quantitative methods such as the PSNR, average gradient, and MSE. The PSNR is calculated from the measurement of antinoise performance of algorithms and is directly proportional to antinoise performance. The average gradient denotes minor dissimilarities between pixels and is directly proportional to clearer image. MSE measures the average of squares of errors that is the average squared change between the expected output and desired output.[7,8]

7.3.2 SKULL STRIPPING

The next step of segmentation process is to perform skull stripping on brain MRI. This process includes removing all the nonbrain tissues such as skull, fat tissues, or cerebral tissues. There are several algorithms proposed for skull stripping such as automatic image contour-based, morphological operation-based, and histogram analysis-based skull stripping or a threshold value.[6,7]

7.3.3 FEATURE EXTRACTION

In this step, all the high-level features such as shape, contrast, texture, and color from the preprocessed and skull-stripped brain images are extracted. This results in efficient and accurate segmentation of the images.

7.3.4 SEGMENTATION

Segmentation is a process in which all the major components of brain images such as GM, WM, CSF, and lesions or tumors are segmented. Various methods have been researched upon for segmentation of brain

MRI such as thresholding methods, region-based methods, and atlas-based methods. Recent research for segmenting brain tumors from MRI is focused on automatic segmentation methods such as deep neural networks, CNNs, DBN, RBM, and stacked auto-encoder networks (SAEs) that are being classified in Figure 7.2.

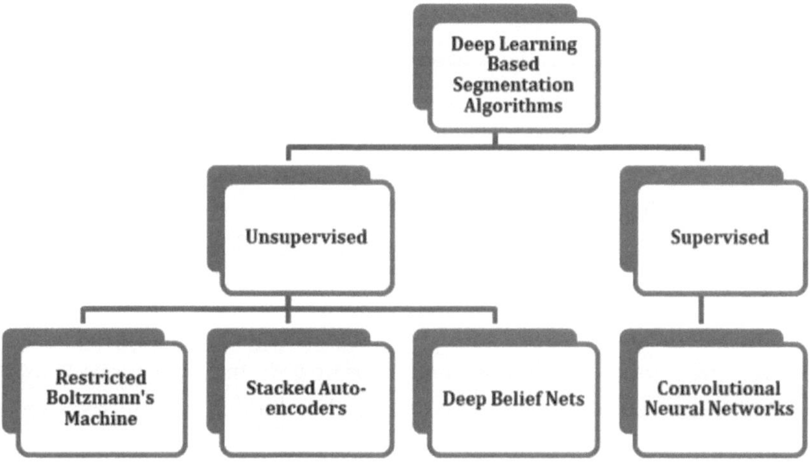

FIGURE 7.2 Deep learning-based segmentation algorithms.

7.4 DEEP LEARNING TECHNIQUES

Immense research is going on toward the automation of medical image analysis that includes segmentation, feature extraction, image registration, and reconstruction. Earlier traditional methods used for segmentation, dealt with certain issues such as:

- Features were extracted manually in traditional methods to pass as the input to the segmentation process.
- Segmentation was difficult to perform on input images with higher dimensions.
- Classification was performed after segmentation.
- These methods require extensive preprocessing techniques prior to segmentation.

To overcome the limitations of traditional segmentation methods, research is accelerated toward deep learning techniques for segmentation process. Recent deep learning techniques based on creating illustration or abstraction at several levels have been explored for the segmentation of tumors from brain MRI. Deep learning algorithms for brain tumor segmentation classified as unsupervised and supervised techniques. Unsupervised techniques include RBM, SAEs, and deep belief nets. Supervised techniques include CNNs.

7.4.1 UNSUPERVISED LEARNING

Unsupervised learning algorithms take input images without the labeled data and are trained to find various similar patterns. These tasks are performed under different loss functions.[9]

7.4.1.1 RESTRICTED BOLTZMANN'S MACHINE

RBMs are a form of Markov random fields, comprising input layer or visible layer and a hidden layer that fetches hidden feature representation. There are bidirectional connections between the nodes, so latent feature representation extracted from an input vector, and vice versa. It is a generative model as the new data points are generated.[10] An energy function defined for a specific state of input *i* and hidden units *h*.

Equation 7.1: Energy function.

$$E(i,h) = h^T Wi - c^T i + B^T i \tag{7.1}$$

where c and B are bias terms, and W is the weight matrix. This energy is exponentially distributed and normalized for computing the probability $p(i,h)$ of the system's state.

Equation 7.2: Probability of system's state.

$$p(i,h) = \frac{1}{P} \exp\{-E(i,h)\} \tag{7.2}$$

Figuring the partition function P is commonly fixed. However, restricted inference of calculating trained on or vice versa is flexible and results as follows:

Equation 73: Conditional inference.

$$P(h_j|i) = \frac{1}{1+\exp\{-b_j - W_j i\}} \tag{7.3}$$

7.4.1.2 STACKED AUTO-ENCODER NETWORK

Stacked auto-encoders are deeper networks consisting of stacked layers of denoising auto-encoders. Auto-encoders are trained to reconstruct a repaired input from a corrupted version of input. Auto-encoder layers are trained independently and finetuned using supervised training. Denoising auto-encoders are minimizing the loss from the original image to the reconstructed image.[11] AEs are simple networks in which the input is reconstructed over the output layer on one hidden layer. These networks are fixed by a weight matrix W as well as bias B extracted from input along with parallel bias from hidden layer processed to hidden state for the reconstruction. Hidden activation is calculated by a nonlinear function:

Equation 7.4: Hidden activation function.

$$H_l = \sigma(W_{i,h}i + B_{i,h}) \tag{7.4}$$

7.4.1.3 DEEP BELIEF NETWORKS

DBN diverge from stacked auto-encoders as the auto-encoder layers of SAE are replaced by RBMs in DBNs. The belief nets neurons are densely connected for rapid and accurate learning of a good set of parameters. However, learning is difficult in belief nets with many hidden layers; to overcome this scenario, Hinton[12] proposed a firm and greedy algorithm for deep belief nets. In this algorithm, the top two layers act as associative memory and next hidden layers form an acyclic graph, which fetches information from associative memory and converts it into variables such as pixels of the image.

7.4.2 SUPERVISED LEARNING

Supervised learning algorithms take two types of input values: input features and labeled data pairs. These labeled data pairs can appear in several forms such as scalar or vector based on the problem to be solved.

7.4.2.1 CONVOLUTIONAL NEURAL NETWORK

CNNs are deep learning algorithms that take input as an image. CNN consists of convolutional layer, pooling layer, dropout layer, and fully connected layer. The convolution operation performed with a set of *N* kernels with weights and added biases individually creating a new feature map on the input image at each layer. These features forced to an element-by-element nonlinear transform for every convolutional layer *l*:

Equation 7.5: Nonlinear transform.

$$\mathrm{F}_N^l = \sigma(W_N^{l-1} * F^{l-1} + B_N^{l-1}) \tag{7.5}$$

The pooling layer performs downsampling by reducing the dimensions of data by combining output of neuron clusters at one layer into single neuron cluster in another layer. The dropout layer prevents overfitting, which occurs due to excess of parameters in the network. At training stage, certain nodes are either dropped out or kept based on their probability. The fully connected layer receives the output from previous layers and performs high-level reasoning in the network.

7.5 REVIEW OF LITERATURE

According to the literature survey, CNNs have given benchmark results for medical image segmentation. They have different advantages over other artificial neural networks as follows:

- The complete image or a distinct image slice is processed as input.
- It consists of deeper networks with multiple layers.
- Feature extraction is done automatically.

Many researchers have focused on different CNN architectures to improve the brain tumor segmentation. A six-layered 3D CNN with local and global approximations has been proposed by Choi and Jin for striatum segmentation.[13] This model improved segmentation results as dropout function applied in last convolutional layer that reduce overfitting and learn features that are more robust. In this, T1-weighted MRI images are fed to the global CNN, which determines the approximate location of stratum. The volume of striatum is extracted from global CNN and fed to the local CNN, which predicts the accurate label of all boxes. The proposed algorithm obtained higher dice similarity coefficient and precision score.

A multimodal 2D CNN is for MRI segmentation in isointense stage where T1 and T2 images are fed as input to the CNN proposed by Zhang et al.[14] and segmentation maps are received as output. Through intermediate layers of CNN that are convolution and pooling, other operations such as dropout function highly nonlinear mappings are captured. This method gives better results for infant brain segmentation.

A multiscale late fusion CNN with convolution layer, rectified linear unit, and max pooling layer proposed by Bao and Chung.[15] The convolution layer comprises two consecutive operations, that is, linearly transform and nonactivation function executed for feature map generation. Since multiscale CNN was unable to smoothly capture discriminative features from MRI due to multifaceted background in brain images, a novel label consistency method under the framework of the random walker was enhanced. This method led to better and efficient segmentation quality.

A deep voxel-wise residual network proposed for tissue segmentation that combine features from different layers.[16] This model is developed initially deep residual learning for 2D image recognition and further enhanced with 3D variant for handling volumetric data. It also incorporates the appearance features of an image at a low level, vital outline information and elevated context together for refining the volumetric segmentation efficiency.

A multiscale CNN for lesion segmentation on nonuniformly experimented area to incorporate a deeper framework with a foviation effect was proposed by Ghafooriyan et al.[17,18] and it also adds anatomical location information into the network. This method segmented WM intensities. In continuation to this work, FCN with 3D CNN for reducing false positives was proposed for candidate segmentation for lacunae detection.[17] This method is further equipped with contextual information using multiscale analysis and combination of explicit location features.

A two-pathway CNN with different receptive fields adapts both local and global features and shifts abstraction layer that converts feature maps to their data with missing modalities.[3] The proposed network was 30 times better due to its flexible conduct. Kleeseik et al.[19] proposed 3D fully CNN for brain extraction on multimodal input both for enhanced contrast and nonenhanced contrast images, hence providing the average dice score, high specificity, and average sensitivity.

Milletari et al.[20] proposed hough-voting to plot from CNN structures to full patch segmentations. This approach is more robust, flexible, and efficient for multimodal segmentation as it integrates the abstraction capabilities of CNN. An methodology toward 3D volumetric medical image segmentation was proposed for learning to predict the segmentation of whole volume at once.[21]

Moeskops et al.[22] proposed CNN trained on multiple patch sizes. In this, dice coefficient-based innovative objective function enhanced during training to deal with the strong differences between the frontend and backend intensity values of the image voxels. Pereira et al.[23] proposed CNN-based tumor segmentation of input from multiple modalities. This method acquires multiscale information about each voxel by using multiple patch and kernel sizes. It is independent on explicit features but acquires features, which are necessary for classification. Shakeri et al.[24] proposed fully convolutional network trailed by Markov random fields. Alpha-expansion used to achieve imprecise inference imposing dimensional volumetric homogeneity to the CNN.

Zhao and Jia[25] proposed multiscale CNN tumor segmentation with late fusion architecture for pixel-wise classification. This information is integrated from top-three scales of image sizes which are detected by both global and local features of pixels. Pan et al.[26] proposed 2D tumor patch classification using a CNN in which images are directly fed to the CNN which selects the features with self-learning of the network. Dou et al.[27,28] proposed 3D FCN for candidate segmentation and microbleed detection trailed by another 3D CNN for reduction in false positives. It utilizes spatial contextual information completely to extract high-level features of cerebral microbleeds and hence achieve much better detection accuracy.

Segnet, a deep neural network, for automatic segmentation of brain MRI, allocates each voxel to its equivalent anatomical region in an MR image of the brain.[29] Around the different voxels of interest, the information is captured at the inputs of the network, that is around three-dimensional and orthogonal two-dimensional intensity patches seizure a local spatial framework while downscale large two-dimensional orthogonal patches and spaces to the local centroids impose global dimensional consistency.

A deep convolutional encoder neural network for lesion segmentation from brain MRI proposed to integrate convolutional and deconvolutional layers to extract features and segmentation prediction in a single model.[30] This model speeds up the training by acquiring features from entire images, which eradicates patch selection and redundant calculations at the intersection of neighboring patches.

Konstantinos et al. introduced 13D dual pathway CNN which was 11-layer deep, for the separation of brain lesions.[31] When multimodal 3D patch processes at multiple scales within the developed system, it fragments voxel-wise pathology. This network processes a 3D brain volume in 3 min.

Konstantinos et al. proposed 3D 11-layer deep fully connected CCN with conditional random field which adapts to class imbalance and is able to recognize the image regions of lesions.[32] This method is applied on three challenging tasks of lesion segmentation in ischemic stroke, traumatic injuries, and brain tumors. Since this method is computationally effective, it is adopted in a variety of research fields.

Mark Lyksborg et al.[33] proposed a combination of 2D CNNs for execution of a volumetric segmentation on magnetic resonance images. It comprises three networks, skilled from three orthogonal planes. First, the full tumor region segmented from the background. After this, the segmentation is refined using the grow cut method (cellular automaton-based seed growing method). To conclude, subregions within the tumor are segmented using an added group of networks trained for the task. Liyue Shen et al.[34] proposed three CNN-based architectures, namely, baseline voxel-wise CNN, fully convolutional patch-wise CNN, and full image fully convolutional CNN for glioma segmentation. The first two architectures performed well with a dice score of 0.84 and 0.86, respectively, as compared to the third architecture. DNNs have recognized to be the recent technology in brain image analysis; deep CNNs have given the benchmark results for all the challenges including brain image analysis such as BRATS. According to Table 7.1, deep learning algorithms compared based on following parameters such as computation time, dice score, precision, sensitivity, recall, and specificity.

7.6 CONCLUSION AND FUTURE SCOPE

Deep learning has proven to be advanced for automated brain tumor segmentation. According to the literature survey, the recently researched field for brain tumor segmentation is CNNs as volumetric medical images are fed directly as the input and features are extracted automatically by the trained CNN. These advantages have given more accurate and sensitive segmentation results. Hence, CNNs have overcome traditional methods of medical image segmentation. CNNs along with its advantages also come with certain challenges. CNNs require a large amount of dataset, which is an obstacle. CNNs require a manually labeled training dataset, which is a laborious task. CNNs with more number of hidden layers have given promising results. However, this raises space complexity that could be

TABLE 7.1 Deep Learning Algorithms[35] (According to the Literature Survey) and Their Reported Results.

Technique	Image modality	Computation time	Dice score	Precision	Sensitivity	Recall	Specificity
Choi[13]	T1	3 s GPU 1.5 min CPU	0.826 ± 0.038	0.917 ± 0.028		0.756 ± 0.066	
Zhang[14]	T1 and T2		3.30E – 03				
Bao[15]	T1		0.850				
Chen[16]	T1-weighted, T1-IR and T2-FLAIR		86.12				
Ghafooriyan[18] (2016)	FLAIR		0.791				
Ghafooriyan[17] (2017)	T1 and FLAIR				0.974		
Havaei[3]	T1, T2, T1C, and FLAIR	30 times faster	0.85		0.87		0.89
Milletari[21]			0.83		0.82		0.79
Milletari[20]			0.869 ± 0.033				
Moeskops[22]	T1		0.7353				
Pan[26]					Intersection 0.6667		
Pinto[36]			0.78				
Shakeri[24]	T1-weighted		0.87				
Zhao[25]	T1, T1-enhanced, T2, and FLAIR		Accuracy = 0.81	Variance = 0.99			

TABLE 7.1 *(Continued)*

Technique	Image modality	Computation time	Dice score	Precision	Sensitivity	Recall	Specificity
Brebesson[29]			0.725				
Konstantinos[31]	MPRAGE, axial FLAIR, T2		64%				
Konstantinos[32]	MPRAGE, axial FLAIR, T2		89.8	91.5	89.1		
Liyue Shen[34]			BCN = 0.84 FCN = 0.86				

reduced in further research. Deep CNNs are also improvised by incorporating geometric constraints into segmentation models. Self-learning property of CNN architectures can be further utilized to improve the segmentation accurateness by extracting richer peripheral information and selective fuzzy points.

The biggest challenge in automated brain tumor segmentation is to accurately localize the tumor and predict the tumor growth. CNN architecture designed should be more robust and efficient for all the image modalities. Segmentation results could be improvised by using efficient preprocessing techniques. Small transformations like rotations, scaling, and noise specifically possible in real MRIs if integrated with CNNs which improve medical image segmentation results.

KEYWORDS

- **restricted Boltzmann machine**
- **deep learning**
- **convolutional neural network**
- **deep belief nets**
- **stacked auto-encoder**

REFERENCES

1. Bengio, Y.; Lamblin, P.; Popovici, D.; Larochelle, H. Greedy Layer-Wise Training of Deep Networks. *Adv. Neural Inf. Process. Syst.* **2007,** *19* (1), 153.
2. Bahadure, N. B.; Ray, A. K.; Thethi, H. P. Image Analysis for MRI Based Brain Tumor Detection and Feature Extraction Using Biologically Inspired BWT and SVM. *Int. J. Biomed. Imaging* **2017,** *2017*, 1–13.
3. Havaei, M.; Davy, A.; Warde-Farley, D.; Biard, A.; Courville, A.; Bengio, Y.; Pal, C.; Jodoin, P. M.; Larochelle, H. Brain Tumor Segmentation with Deep Neural Networks. *Med. Image Anal.* **2017,** *35*, 18–31.
4. Agravat, R. R.; Raval, M. S. *Deep Learning for Automated Brain Tumor Segmentation in MRI Images*; Elsevier Inc.: Amsterdam, 2018.
5. Roy, P.; Goswami, S.; Chakraborty, S.; Azar, A. T.; Dey, N. Image Segmentation Using Rough Set Theory. *Int. J. Rough Sets Data Anal.* **2014,** *1* (2), 62–74.

6. Benson, C. C.; Lajish, V. L. Morphology Based Enhancement and Skull Stripping of MRI Brain Images. In *2014 Int. Conf. Intell. Comput. Appl. ICICA 2014*, 2014; pp 254–257.
7. Chen, C.-M.; Chen, C.-C.; Wu, M.-C.; Horng, G.; Wu, H.-C.; Hsueh, S.-H.; Ho, H.-Y. Automatic Contrast Enhancement of Brain MR Images Using Hierarchical Correlation Histogram Analysis. *J. Med. Biol. Eng.*, **2015,** *35* (6), 724–734.
8. Isa, I. S.; Sulaiman, S. N.; Mustapha, M.; Karim, N. K. Automatic Contrast Enhancement of Brain MR Images Using Average Intensity Replacement Based on Adaptive Histogram Equalization (AIR-AHE). *Biocybern. Biomed. Eng.* **2017,** *37* (1), 24–34.
9. Litjens, G.; Kooi, T.; Bejnordi, B. E.; Setio, A. A. A.; Ciompi, F.; Ghafoorian, M.; van der Laak, J. A. W. M.; van Ginneken, B.; Sánchez, C. I. A Survey on Deep Learning in Medical Image Analysis. *Med. Image Anal.* **2017,** *42* (December 2012), 60–88.
10. Hinton, G.; Hinton, G. A Practical Guide to Training Restricted Boltzmann Machines. In *Neural Networks: Tricks of the Trade. Lecture Notes in Computer Science*; Montavon, G., Orr, G. B., Müller, K. R. (Eds.); Springer: Berlin, Heidelberg, 2010; Vol 7700.
11. Vincent, P.; Larochelle, H. Stacked Denoising Autoencoders: Learning Useful Representations in a Deep Network with a Local Denoising Criterion Pierre-Antoine Manzagol. *J. Mach. Learn. Res.* **2010,** *11*, 3371–3408.
12. Hinton, G. E.; Osindero, S.; Teh, Y.-W. A Fast Learning Algorithm for Deep Belief Nets. *Neural Comput.* **2006,** *18* (7), 1527–1554.
13. Choi, H.; Jin, K. H. Fast and Robust Segmentation of the Striatum Using Deep Convolutional Neural Networks. *J. Neurosci. Methods* **2016,** *274*, 146–153.
14. Zhang, W.; Li, R.; Deng, H.; Wang, L.; Lin, W.; Ji, S.; Shen, D. Deep Convolutional Neural Networks for Multi-Modality Isointense Infant Brain Image Segmentation. *NeuroImage* **2015,** *108*, 214–224.
15. Bao, S.; Chung, A. C. S. Multi-Scale Structured CNN with Label Consistency for Brain MR Image Segmentation. *Comput. Methods Biomech. Biomed. Eng. Imaging Vis.* **2018,** *6* (1), 113–117.
16. Chen, H.; Dou, Q.; Yu, L.; Qin, J.; Heng, P. A. VoxResNet: Deep Voxelwise Residual Networks for Brain Segmentation from 3D MR Images. *NeuroImage* **2017,** *170*, 446–455.
17. Ghafoorian, M.; Karssemeijer, N.; Heskes, T.; Bergkamp, M.; Wissink, J.; Obels, J.; Keizer, K.; de Leeuw, F. E.; van Ginneken, B.; Marchiori, E.; Platel, B. Deep Multi-scale Location-Aware 3D Convolutional Neural Networks for Automated Detection of Lacunes of Presumed Vascular Origin. *NeuroImage Clin.* **2017,** *14*, 391–399.
18. Ghafoorian, M.; Karssemeijer, N.; Heskes, T.; van Uden, I.; Sanchez, C.; Litjens, G.; de Leeuw, F.-E.; van Ginneken, B.; Marchiori, E.; Platel, B. Location Sensitive Deep Convolutional Neural Networks for Segmentation of White Matter Hyperintensities. *Sci. Rep.* **2016,** *5110* (2017). https://doi.org/10.1038/s41598-017-05300-5.
19. Kleesiek, J.; Urban, G.; Hubert, A.; Schwarz, D.; Maier-Hein, K.; Bendszus, M.; Biller, A. Deep MRI Brain Extraction: A 3D Convolutional Neural Network for Skull Stripping. *NeuroImage* **2016,** *129*, 460–469.
20. Milletari, F.; Ahmadi, S. A.; Kroll, C.; Plate, A.; Rozanski, V.; Maiostre, J.; Levin, J.; Dietrich, O.; Ertl-Wagner, B.; Bötzel, K.; Navab, N. Hough-CNN: Deep Learning for

Segmentation of Deep Brain Regions in MRI and Ultrasound. *Comput. Vis. Image Understand.* **2017,** *164*, 92–102.

21. Milletari, F.; Navab, N.; Ahmadi, S.-A. V-Net: Fully Convolutional Neural Networks for Volumetric Medical Image Segmentation. *Comput. Vis. Pattern Recogn.* **2016,** 1–11.
22. Moeskops, P.; Viergever, M. A.; Mendrik, A. M.; De Vries, L. S.; Benders, M. J. N. L.; Isgum, I. Automatic Segmentation of MR Brain Images with a Convolutional Neural Network. *IEEE Trans. Med. Imaging* **2016,** *35* (5), 1252–1261.
23. Menze, B. H.; Jakab, A.; Bauer, S.; Kalpathy-Cramer, J.; Farahani, K.; Kirby, J.; Burren, Y.; Porz, N.; Slotboom, J.; Wiest, R.; Lanczi, L.; Gerstner, E.; Weber, M. A.; Arbel, T.; Avants, B. B.; Ayache, N.; Buendia, P.; Collins, D. L.; Cordier, N.; Corso, J. J.; Criminisi, A.; Das, T.; Delingette, H.; Demiralp, Ç.; Durst, C. R., Dojat, M.; Doyle, S.; Festa, J.; Forbes, F.; Geremia, E.; Glocker, B.; Golland, P.; Guo, X.; Hamamci, A.; Iftekharuddin, K. M.; Jena, R.; John, N. M.; Konukoglu, E.; Lashkari, D.; Mariz, J. A.; Meier, R.; Pereira, S.; Precup, D.; Price, S. J.; Raviv, T. R.; Reza, S. M. S.; Ryan, M.; Sarikaya, D.; Schwartz, L.; Shin, H. C.; Shotton, J.; Silva, C. A.; Sousa, N.; Subbanna, N. K.; Szekely, G.; Taylor, T. J.; Thomas, O. M.; Tustison, N. J.; Unal, G.; Vasseur, F.; Wintermark, M.; Ye, D. H.; Zhao, L.; Zhao, B.; Zikic, D.; Prastawa, M.; Reyes, M.; Van Leemput, K. The Multimodal Brain Tumor Image Segmentation Benchmark (BRATS). *IEEE Trans. Med. Imaging* **2015,** *34* (10), 1993–2024.
24. Shakeri, M.; Tsogkas, S.; Ferrante, E.; Lippe, S.; Kadoury, S.; Paragios, N.; Kokkinos, I.; Paris-Saclay, U.; Montreal, P. Sub-Cortical Brain Structure Segmentation Using F-CNN'S University of Montreal, 4 Sainte-Justine Hospital Research Center. *Isbi* **2016,** *2016*, 269–272.
25. Zhao, L.; Jia, K. Multiscale CNNs for Brain Tumor Segmentation and Diagnosis. *Comput. Math. Methods Med.* **2016,** *2016.*
26. Pan, Y.; Huang, W.; Lin, Z.; Zhu, W.; Zhou, J.; Wong, J.; Ding, Z. Brain Tumor Grading Based on Neural Networks and Convolutional Neural Networks. In *Eng. Med. Biol. Soc. (EMBC), 2015 37th Annu. Int. Conf. IEEE*, 2015; pp 699–702.
27. Dou, Q.; Chen, H.; Yu, L.; Shi, L.; Wang, D.; Mok, V. C.; Heng, P. A. Automatic Cerebral Microbleeds Detection from MR Images Via Independent Subspace Analysis Based Hierarchical Features. In *Proc. Annu. Int. Conf. IEEE Eng. Med. Biol. Soc. EMBS*, 2015, vol. 2015, Novem.; pp 7933–7936.
28. Dou, Q.; Chen, H.; Yu, L.; Zhao, L.; Qin, J.; Wang, D.; Mok, V. C. T.; Shi, L.; Heng, P. A. Automatic Detection of Cerebral Microbleeds from MR Images Via 3D Convolutional Neural Networks. *IEEE Trans. Med. Imaging* **2016,** *35* (5), 1182–1195.
29. de Brebisson, A.; Montana, G. Deep Neural Networks for Anatomical Brain Segmentation. *Comput. Vis. Pattern Recogn.* **2015,** 20–28.
30. Brosch, T.; Yoo, Y.; Tang, L. Y. W.; Li, D. K. B.; Traboulsee, A.; Tam, R. Deep Convolutional Encoder Networks for Multiple Sclerosis Lesion Segmentation. In Medical Image Computing and Computer-Assisted Intervention—MICCAI 2015. MICCAI 2015. Lecture Notes in Computer Science; Navab, N., Hornegger, J., Wells, W., Frangi, A. (Eds.); Springer, Cham, 2015; Vol 9351. https://doi.org/10.1007/978-3-319-24574-4_1.
31. Kamnitsas, K.; Chen, L.; Ledig, C.; Rueckert, D. Efficient Multi-Scale 3D Convolutional Neural Networks for Lesion Segmentation in Brain MRI. In *Proc. MICCAI Ischemic Stroke Lesion Segmentation Challenge*, 2015; pp 7–10.

32. Kamnitsas, K.; Ledig, C.; Newcombe, V. F. J.; Simpson, J. P.; Kane, A. D.; Menon, D. K.; Rueckert, D.; Glocker, B. Efficient Multi-scale 3D CNN with Fully Connected CRF for Accurate Brain Lesion Segmentation. *Med. Image Anal.* **2017,** *36*, 61–78.
33. Lyksborg, M.; Puonti, O.; Agn, M.; Larsen, R. An Ensemble of 2D Convolutional Neural Networks for Tumor Segmentation. In *Image Analysis. SCIA 2015. Lecture Notes in Computer Science*; Paulsen, R., Pedersen, K. (Eds.); Springer: Cham, vol 9127. https://doi.org/10.1007/978-3-319-19665-7_17.
34. Shen, L.; Anderson, T. Multimodal Brain MRI Tumor Segmentation via Convolutional Neural Networks, **2015,** *18* (5), 2014.
35. Sille, R.; Chauhan, P.; Sharma, D. Deep Learning based Brain MRI Segmentation Algorithms, 2019.
36. Pinto, A.; Alves, V.; Silva, C. A. Brain Tumor Segmentation Using Convolutional Neural Networks in MRI Images. *IEEE Trans. Med. Imaging* **2016,** *35* (5), 1240–1251.

CHAPTER 8

Recurrent Neural Networks and Their Application in Seizure Classification

KUSUMIKA KRORI DUTTA[1], POORNIMA SRIDHARAN[2], and SUNNY AROKIA SWAMY BELLARY[3]

[1]*Assistant Professor, M.S. Ramaiah Institute of Technology, MSR Nagar, Bengaluru, Karnataka 560054, India*

[2]*Research Scholar, Anna University, India Chennai, Tamil Nadu 600025, India*

[3]*IEEE Member, Charlotte, NC 28262, USA*

[*]Corresponding author. E-mail: kusumika@msrit.edu

ABSTRACT

Deep learning (DL) architectures such as deep neural networks (DNN), deep belief networks (DBN), recurrent neural networks(RNN) and convolutional neural networks (CNN) have been applied to applications such as computer vision, speech recognition, natural language processing, audio recognition, social network filtering, machine translation, bioinformatics, drug design, medical image analysis, material inspection and board game programs, in which has comparable performance than human experts. With the growing interest and research in the area of artificial neural network, deep neural network enable computers to get trained for error-free diagnosis to diseases like epilepsy. In literature, researchers carried out many mathematical models for pre-processing of EEG data and classification between seizure and seizure free signals or different

Deep Learning in Visual Computing and Signal Processing. Krishna Kant Singh, Vibhav Kumar Sachan, Akansha Singh & Sanjeevikumar Padmanaban (Eds.)

types of network disorders. The introduction of various algorithms like machine learning deep learning, etc., in artificial intelligence, aids to classify the data with or without pre-processing and two class system. It is important to try multi-class time series classification of various brain activities (tumour, network disorders) using the sophisticated algorithms.

In this chapter, different deep learning algorithms for multiclass, time series classification of different electrical activities in brain are discussed. The main focus is on the application of different RNN models in seizure classification of Electroencephalogram (EEG) signals. It is very important to interpret the 1D EEG signals and classify among different activities of brain for various diagnostic purpose. The fully interconnected hidden configuration of recurrent neural network (RNN) makes the model very dominant which enable to discover temporal correlations between far away events in the data. The training of RNN architecture when used in deep network is challenging because of vanishing/exploding gradient in deeper layer. This paper aims to perform multiclass time series classification of EEG signal using three different RNN techniques; simple Recurrent Neural Network, Long-Short Term Memory (LSTM) and GRUs. A comparative study between RNNs is done in terms of configuration, time taken and accuracy for EEG signals acquired from people having different pathological and physiological brain states. The accuracy and time taken for multilayer recurrent neural networks are determined for classification of EEG for five different classes using three different types of RNN networks, for 1 to 1024 units with 100 epochs and 5 different layers of 32 cells with 300 epochs, with a learning rate of 0.01. It has been observed that the number of layers increases the time complexity and provides constant accuracy for more than three layers. Further, it can be extended for the accuracy and time consumption for different batch sizes with different epochs to fix a proper network without over fitting the network.

8.1 RECURRENT NEURAL NETWORK

A grand intelligence of artificial neural networks (ANN) is used to compute the learning models of human nervous systems. The development of ANN in various applications such as image recognition and classification, self-driving vehicles, data analytics and prediction, animation games, and so on, proved that the performance of ANN models is equal or better than that of human. The basic architectures of ANN accept dependent multidimensional data and complicate its structure according to the application requirement. The evolution of recurrent neural networks (RNNs)-based deep learning (DL) architecture innovated many applications in time series, biological, speech to text conversion, which has sequence-dependent data. RNN handles both real values (time series) and symbolic values of variable length inputs. Semantic interpretation of text using time-layered RNN structure found new research areas such as sentiment analysis, machine translation, and information analytics.[9]

RNN holds memory in the form of hidden layers. In RNN, the information is transferred among individual layers of the network and the sequence positions. While the conventional ANN gets complicated with variable length inputs for single layer, RNN accepts single input for multiple layers corresponding to each position of sequence. The input in each layer has similar modeling parameters which repeats the architecture throughout the network and hence "recurrent neural network" defines as a self-loop in the hidden layer neuron enables recurrence of the previous output. The hidden neuron of RNN holds the summary of previous inputs of same sequence, which differentiates it from other feed-forward networks.

8.2 EVOLUTION OF RNN

In 1985, Rumelhart et al.[11] developed a neural network which learns using backpropagation (BPN) of errors and enables easy learning of complex deep networks like restricted Boltzmann machine. Yann Lecun recognized handwritten digits using BPN-based CNN. Sepp Hochreiter found the slow learning of RNN due to vanishing gradient issue in 1991 and discovered long short-term memory (LSTM) neural network in 1997.[44] LSTM allows the error to travel with internal memory through the positions of the sequence. Figure 8.1 shows the development of RNN since its evolution.

Schuster and Paliwal classified phonemes of TIMID dataset by developing a bi-directional RNN in which the neuron's output depends on both the previous and next steps of a sequence. Gers et al. included gates in order to reset the memory on requirement which has become the essential part of vanilla LSTM later. In 2000, Gers and Schmidhuber[15] proposed peephole connections to the neuron of RNN in addition to actual input and proved that time taken to obtain solution is less than LSTM. Graves and

Schmidhuber observed that the learning and accuracy have been improved by applying bi-directional RNN to gated LSTM.

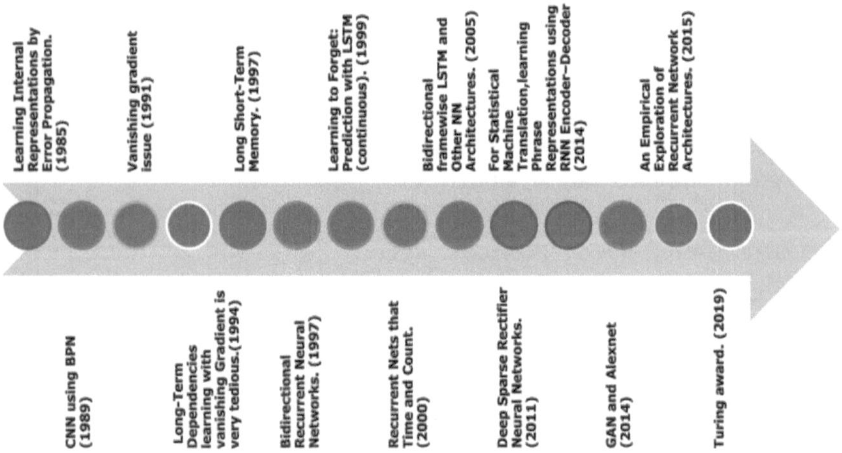

FIGURE 8.1 Evolution of recurrent neural networks

Youshua Bengio[53] discovered that ReLU activation functions could vanishing gradient of RNNs like gated nodes. Cho et al. included reset gate and update gates to RNN nodes to select the nodes to be resetted and generate output. Jozefowicz et al. proved by excluding a bias of a node in LSTM behaves similar to gated recurrent unit (GRU)-based RNN. The contribution of Alex and Ian inspired many researches in the evolution of Alexnet and generative adversarial networks (GANs), respectively, in various field applications.

8.3 ARCHITECTURE OF DIFFERENT RNN

The RNN consists of input, hidden, and output layers. It has recurrent structure between output and input, after completion of each epoch during training of the network. A sequence data with finite length unfold the self-loop of hidden neurons into feed-forward network which resembles the traditional network. The self-loop allows BPN learning within hidden layers for weight updation in discrete steps and employs the same function for each neuron. Figure 8.2 shows the simple architecture of conventional ANN and RNN.

While the term W_h represents the weight connecting the input and hidden layer for an input x which could be one of the position value in a sequence, the weight W_y between hidden and output layer along with activation function generates the output y which could be predicted or matching value of a sequence. The RNN loop can be unfurled as shown in Figure 8.3.

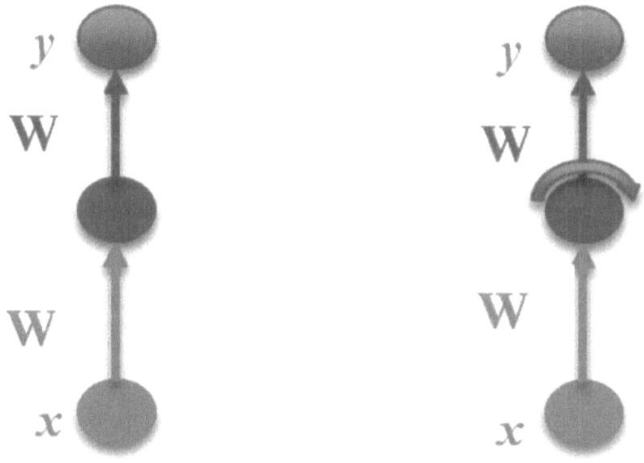

FIGURE 8.2 Basic architecture of ANN and RNN (with loop).

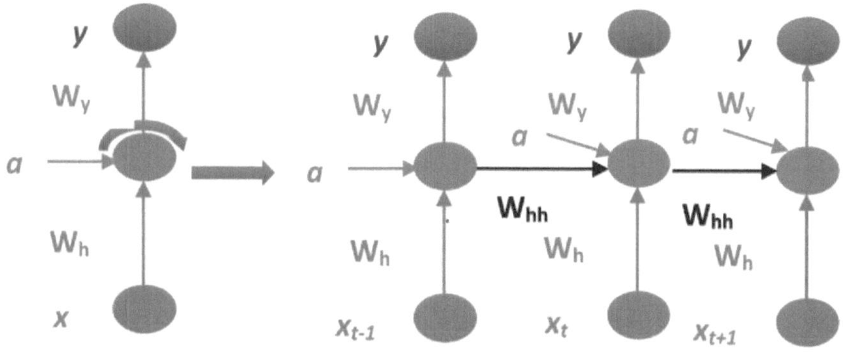

FIGURE 8.3 Basic architecture of ANN and RNN (with loop).

The unfolded structure implies that each neuron in the hidden layer same parameters in which the terms W_{hh} and a indicate the connecting weight between the neurons and activation function, respectively. The weights remain same for all the neurons and shared among them. The network at the input x_t shares information of x_{t-1} through the activation function (a) and weight (W_{hh}) between hidden neurons of x_t and x_{t-1} layers.

The information will be shared among the hidden layers in the forward direction only. The output of hidden neuron is given as

$$H_t = \tanh\left(W_x x_t + W_{hh} x_{t-1}\right) \tag{8.1}$$

The output y_t of network can be written as

$$y_t = W_y H_t \tag{8.2}$$

The output y_t for the input x_t depends on H_t which in turn depends on the inputs x_t and x_{t-1}. Similarly, the output y_{t+1} has same relation with the inputs x_{t+1} and x_t. This method of RNN enables to incorporate inputs with variable lengths (e.g., language modeling). It also works with fixed as well as varying input and output nodes. Figure 8.4 shows the various architecture of RNN being used for practical applications.

The structure of Figure 8.4a has one input and output neuron which suits for image classification. While Figure 8.4b structure has one input and many output neurons, the sequence network belongs to image caption, speech synthesis, music generation, and video games. The structure Figure 8.4c with many input and one output neurons has been used for sentiment analysis. The robotic control, dialogue analysis, machine translation, and video classification sequence network need many input and output neurons as shown in Figure 8.4d. All of the structures are independent on length of input and output neurons due to its recurrence nature. Irrespective of missing input or output nodes, RNN structures dominate other ANN computational models.

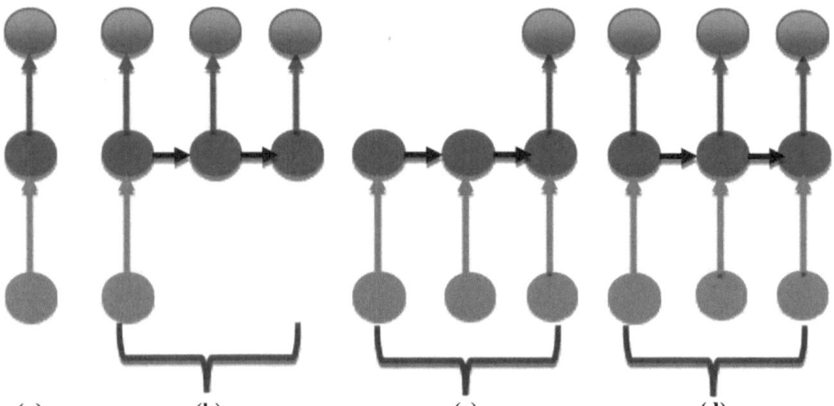

(a) (b) (c) (d)

FIGURE 8.4 Unfolded RNN structure.

The error between expected and actual output is fed back and reduced using BPN by time learning algorithm. These variants enable fast parameter updates and hence the training. Truncated BPN (TBPN) is recommended for the sequence of more than 100 positions. TBPN divides the forward and backward passes into smaller sections which update the parameter simultaneously and reduce the computational effort.

Figure 8.5 shows the example work flow of BPN and TPN for better understanding. If the sequence has more than 10 positions, the memory of first input starts to diminish slowly which is termed as vanishing gradient in RNN. This limitation can overcome by using LSTM and GRU variant-based structures.

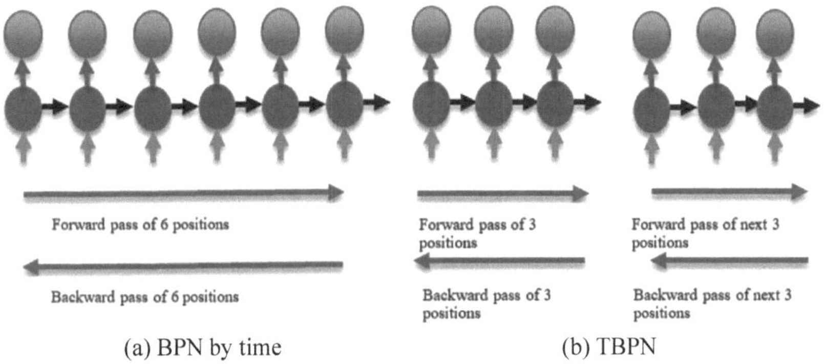

FIGURE 8.5 Workflow of BPN by time and TBPN.

8.4 VARIOUS RNN STRUCTURES

The RNN has various structural modifications to obtain a standard model for specific applications since its origin at 1985. All the variations of RNN architecture have been briefed in Table 8.1. The modification of RNN in terms of structure, learning, and applications is enlisted for easy understanding across the various researches. The structure of each variant has been given according to the literature done.

8.5 TYPES OF RECURRENT NEURAL NETWORK

The types of RNN subdivided based on (1) architecture of RNN and (2) working of RNN. Let us discuss both subdivisions one by one.

TABLE 8.1 Variants of RNN Architecture.

Sl. no.	RNN	Layer structure	Learning	Applications
1	Hopfield[7] 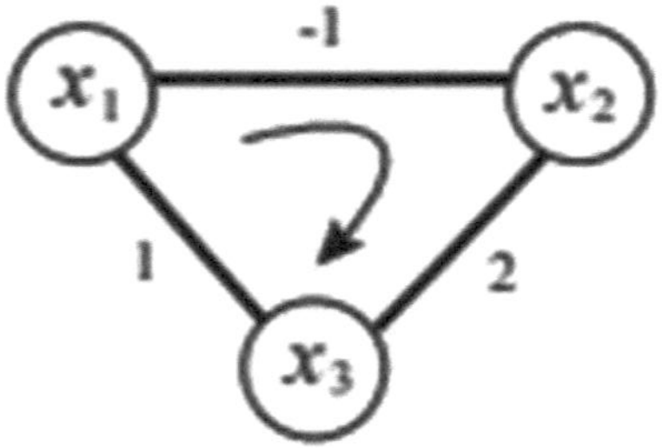 https://machinelearningknowledge.ai/brief-history-of-deep-learning/#Hopfield_Network_8211_Early_RNN	The network does not support sequence inputs but fixed one after finite number of updates. The network enables small memory storage with assured convergence, good efficiency, and flexible property	Hebbian	Image recognition, restoration, enhancement, and detection from highly distorted trained images
2	Echo state[19] 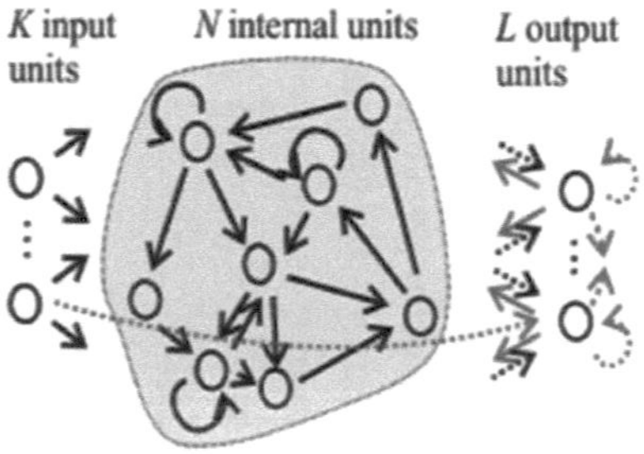	The sequence inputs from input layer fed to a sparsely connected hidden layer. Learning has been applied only to the neurons of output layer	Supervised	Signal processing applications, polymer mixtures, artificial soft limbs, optical microchips, mechanical nano-oscillators

TABLE 8.1 *(Continued)*

Sl. no.	RNN	Layer structure	Learning	Applications
3	Independent RNN[42] 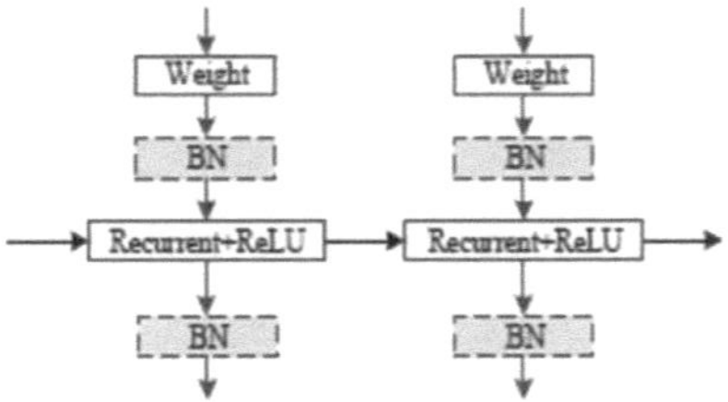	The neurons of same layer are independent with each other though they are connected with neurons of other layers. The nonsaturation activation functions and robust training of neurons aid to outperform over-dependent RNNs or LSTMs. The basic modules are arranged one over the other to frame a deeper structure which can accept up to 5000 time steps	BPN by time	Skeleton-based action recognition, language modeling, sequential MNIST classification
4	Recursive[37] 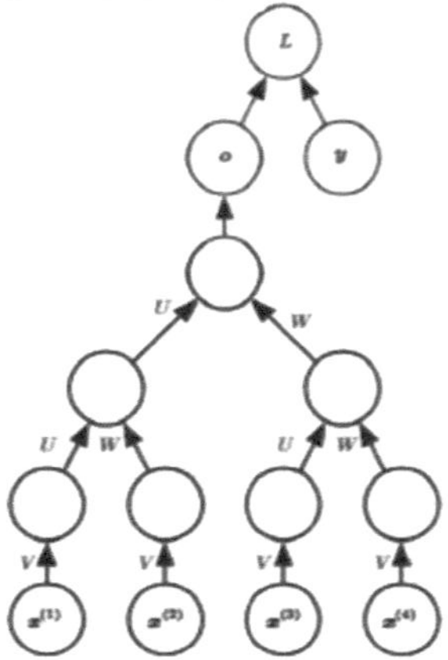	The inputs to recursive network are normally natural scenes or language dataset. The input data have activation vectors representing segments of scene or word of sentence and adjacency matrix which determines the merging of symmetrical neighbor segments. Being recursive, the input of length L could be reduced to $\log(L)$. A fixed output is associated with the variable input sequence using fixed weights of hidden layers which represents a topological graph	BPN	natural language processing, semantic scene segmentation, annotation, and classification

TABLE 8.1 *(Continued)*

Sl. no.	RNN	Layer structure	Learning	Applications
5	Second-order RNN[28] 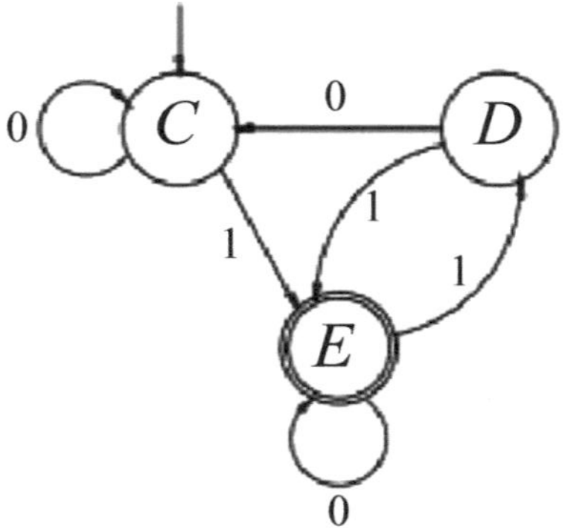	The network uses N^2M weights, whereas single-order network has $N(M + N)$ weights for M inputs and N states. Feed backing several feed-forwarded layers helps to recognize finite-state sequences	Gradient descent	Finite extracted automata classification in compilers, study of language, coin-operated vending machines (turnstiles in amusement parks, metro-subway access)
6	Bidirectional RNN (BRNN)[43] 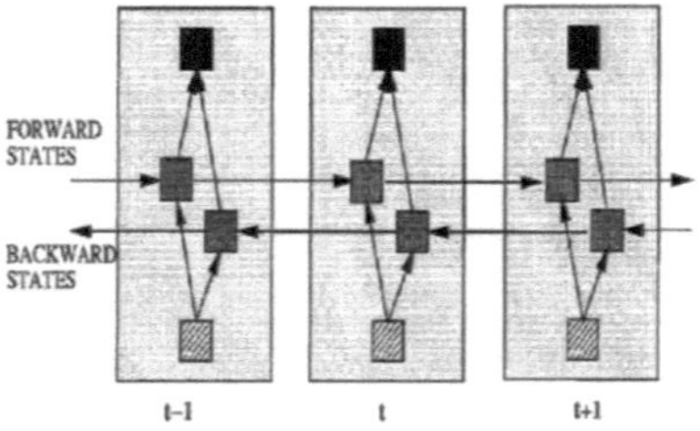	The input layer with finite sequence inputs travels toward the output layer through hidden states which are interconnected in forward and reverse direction through different neurons. None of the outputs are fed back to other directed neurons. Absence of any directed layer resembles single layer RNN in forward or reverse manner. The network outputs the segment with the knowledge of its past and future information	BPN through time	Phoneme classification, speech and handwritten recognition, and protein structure prediction

TABLE 8.1 *(Continued)*

Sl. no.	RNN	Layer structure	Learning	Applications
7	Continuous time (CTRNN)[33] 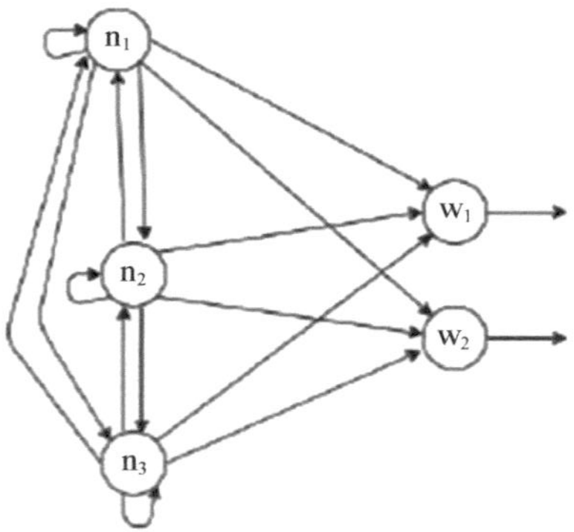	Behaves as a leaky integrator, the network consists of interconnected neurons of hidden layer are self-connected exhibiting recursive and complex dynamics. The internal state is characterized by differential equations in terms of gain, bias and time constant, weight, and nonlinear transfer function. CTRNN outperforms discrete RNN as the output doesn't depend only on *i*th and $(i-1)$th inputs	Evolution algorithms	Evolutionary robotics, beat tracking algorithms, artificial composers, novel musical compositions, autonomous musical software agents, and match an existing piece of audio. Synthesis of the variants of the original sound, audio synthesis
8	Recurrent multilayer perceptron (RMLP)[50] 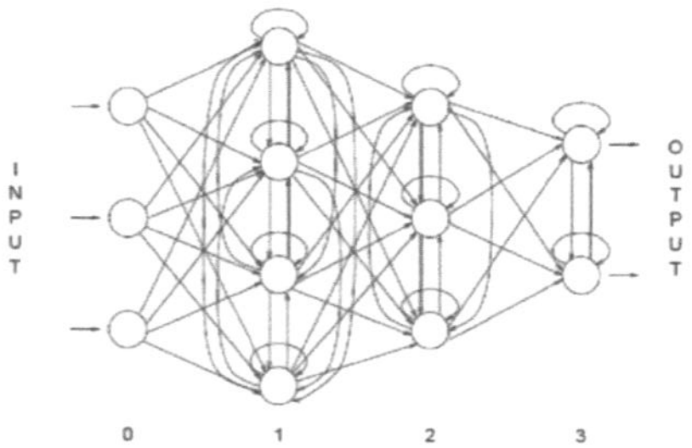	Derived from McCulloch Pitt's model, the network satisfies temporal requirement by recurrence and follow-up of continuity using multifeed forwarded hidden layers. Each hidden layer has interlinked and has its neurons with trained self-links with one-time delay unit. Each hidden layer connected sequentially in forward manner represents a subnet. The total number of subnets in a RMLP is given by sum of hidden subnets and 2	Gradient descent such as BPN through time, dynamic BPN, real-time recurrent learning (RTRL)	Identi fication of automotive environment, connection admission control in ATM network, dynamic channel allocation in mobile cellular networks

TABLE 8.1 *(Continued)*

Sl. no.	RNN	Layer structure	Learning	Applications
9	Multiple timescales RNN[54] 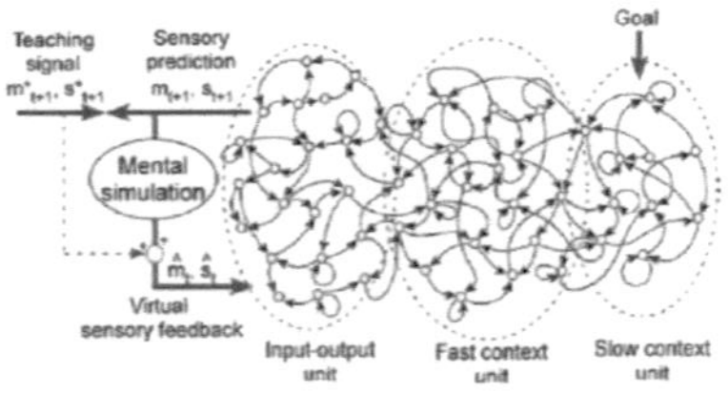	A self-organized CNN for a functional hierarchy has been realized by means of movement of neurons at various timescales. The network consists of input fed-output generated unit and a context unit in which output is generated without input being fed. The context unit consists of fast-acting element with low time constant value and slow-acting element with high time constant value. These elements react to changes based on the respective time constants. In addition, the slow context element is indirectly related to the sequence fed to the layers whose initial state is determined by the expected sequence. Sequential inputs are generated based on the weights of a certain time step	BPTT	Humanoid robot, extraction of the unique slow features from human actions data, recognition of sensory motor patterns

8.5.1 TYPES OF RNN BASED ON ARCHITECTURE

To start with there were only one type of architectures, that is, many to many, and in that case number of input and number of output remains the same. But it has been observed that, in many of the applications, number of input and output varies of each other. So, based on applications, the RNN architectures are (a) one to one, (b) one to many, (c) many to one, (d) many to many (with equal input–output), and (e) many to many (with unequal input–output).

(a) *One to one:* In this, one input and one output is there with no bias as shown in Figure 8.6.

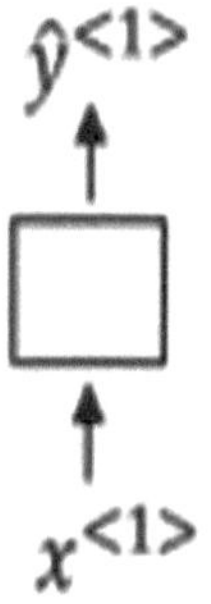

FIGURE 8.6 One to one.

(b) *One to many:* In this case, one input but many output available, as shown in Figure 8.7. This architecture can be applicable to music generation.

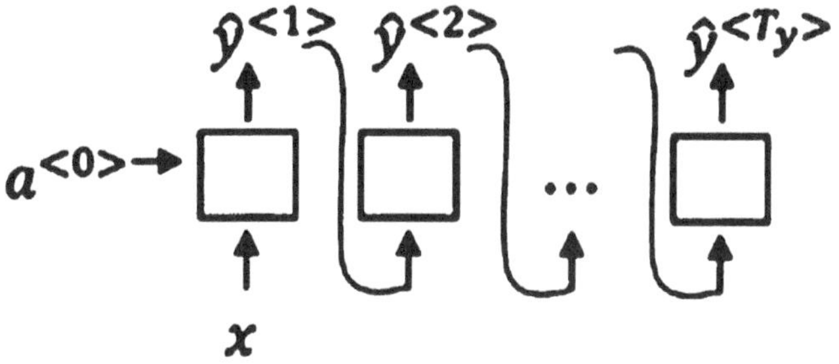

FIGURE 8.7 One to many.

(c) *Many to one:* In this type, many inputs contribute to one output as shown in Figure 8.8. This architecture can be applicable to sentiment analysis.

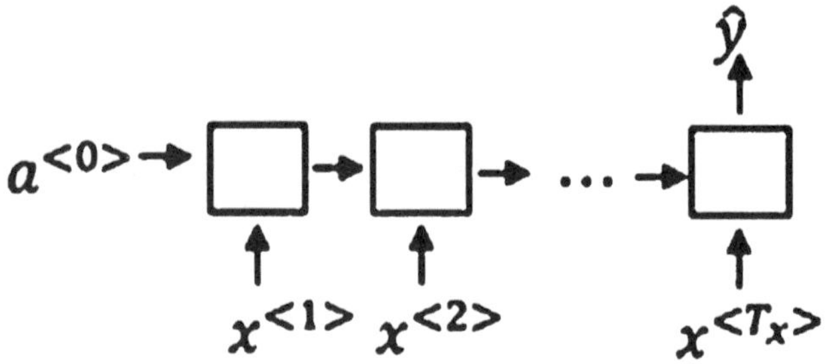

FIGURE 8.8 Many to one.

(d) *Many to many* (with equal input–output): In this case, number of input and number of outputs remain same, as in Figure 8.9. This architecture can be applicable to name entity recognition.

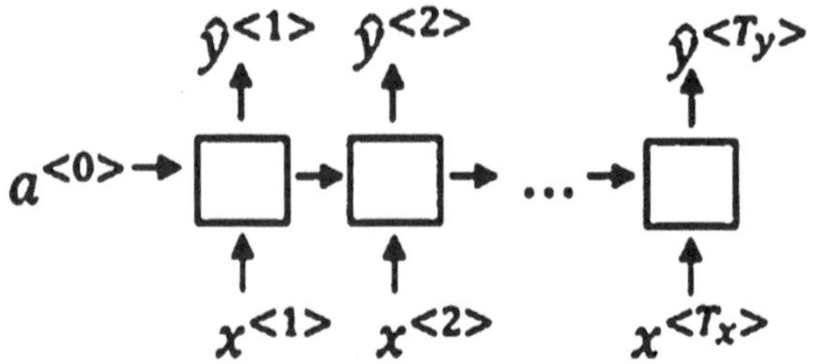

FIGURE 8.9 Many to many, where $T_x = T_y$.

(e) *Many to many (with unequal input and output):* In this case, number of inputs and number of outputs are not same, as in Figure 8.10. This architecture can be applicable to machine translation.

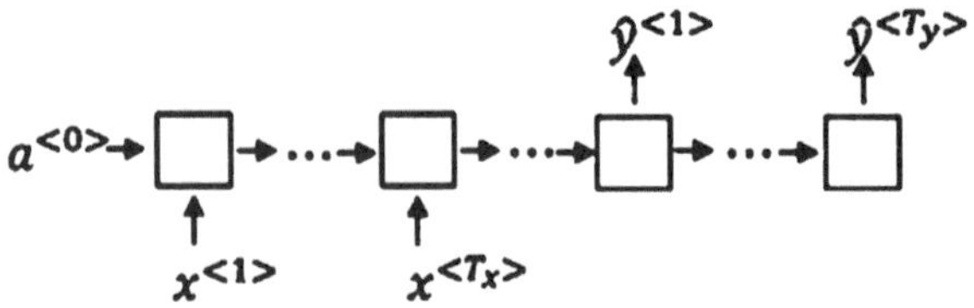

FIGURE 8.10 Many to many where $T_x \neq T_y$.

8.5.2 THE TYPES OF RNN MODEL BASED ON WORKING PRINCIPLE

Based on the working principle, RNNs are of three types: (a) simple RNN, (b) GRUs, and (c) LSTM networks.

8.5.2.1 SIMPLE RECURRENT NEURAL NETWORKS

Simple RNNs have three layers: input, hidden, and output. Based on the architecture simple recurrent network is of two types: (i) Elman networks and (ii) Jordan networks. In case of Elman networks, sets of context units fed from hidden layer are connected, which can store value of previous hidden layer units and sequence prediction. But in case of Jordan networks, output layer feed sets of context units as shown in Figure 8.11.

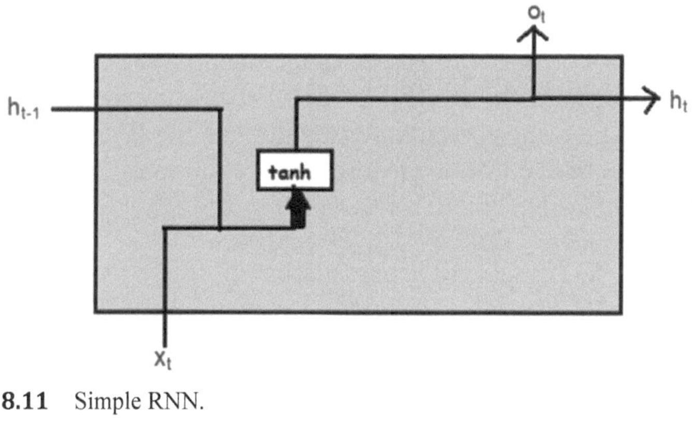

FIGURE 8.11 Simple RNN.

Let us consider
x_t: input vector
h_t: hidden layer vector
o_t: output vector
y_t: target vector
b_x: bias vector (input layer)
b_h: bias vector (hidden layer)
b_o: bias vector (output layer)
W_x, W_h: weight parameter matrix
$e(\,)$, $f(\,)$: a nonlinear activation function (e.g., tan h, sigmoidal)

For total time steps of T:

Feed-forward phase (h_t, y_t, L_t, L): For the given sequence input, the RNN computes the hidden state output and model output as

$$g_t = W_x x_t + W_h h_{t-1} + b_h \tag{8.3}$$

$$h_t = e\left(W_x x_t + W_h h_{t-1} + b_h\right) \tag{8.4}$$

$$z_t = W_h h_t + b_o \tag{8.5}$$

$$o_t = f(h_t) \tag{8.6}$$

The loss function of the network is computed as sum of loss per time-step of the model can be expressed as

$$L\left(o, y\right) = \sum_{t=1}^{T} L(o_t; y_t) \tag{8.7}$$

Bac-propagation phase $\left(\frac{\partial L}{\partial W_o}, \frac{\partial L}{\partial W_h}, \frac{\partial L}{\partial W_x}, \frac{\partial L}{\partial b_o}, \frac{\partial L}{\partial b_h}, \frac{\partial L}{\partial b_x}\right)$: The information of loss function has been passed to the input layer in terms of weight matrices and updates the weights of the network through BPN. All the weights are shared across each time step through the layers of RNN. The gradient descent-based learning algorithm is as follows:

For every time-step "t" from T to 1:

1. $$\frac{\partial L}{\partial W_o} = \sum_{t=1}^{T} \frac{\partial L_t}{\partial W_o}; \frac{\partial L_t}{\partial W_o} = \frac{\partial L_t}{\partial y_t} \cdot \frac{\partial y_t}{\partial W_o} \tag{8.8}$$

   $$\frac{\partial L}{\partial W_h} = \sum_{t=1}^{T} \frac{\partial L_t}{\partial W_h}; \frac{\partial L_t}{\partial W_h} = \frac{\partial L_t}{\partial y_t} \cdot \frac{\partial y_t}{\partial h_t} \cdot \frac{\partial h_t}{\partial W_h} \tag{8.9}$$

2. Differentiating (1.4) and substituting in (1.9)

   $$\frac{\partial L_t}{\partial W_h} = \frac{\partial L_t}{\partial y_t} \cdot \frac{\partial y_t}{\partial h_t}\left(\frac{\partial h_t}{\partial W_h} + \frac{\partial h_t}{\partial h_{t-1}} \cdot \frac{\partial h_{t-1}}{\partial W_h} + \cdots\right) \tag{8.10}$$

3. Also, there are k hidden layers, the above equation can be rewritten as

   $$\frac{\partial L_t}{\partial W_h} = \frac{\partial L_t}{\partial y_t} \cdot \frac{\partial y_t}{\partial h_t} \sum_{k=0}^{t}\left(\frac{\partial h_t}{\partial h_{t-1}} \cdot \frac{\partial h_{t-1}}{\partial W_h} + \cdots \frac{\partial h_{k+1}}{\partial h_k} \cdot \frac{\partial h_k}{\partial W_h}\right) \tag{8.11}$$

4. $$\frac{\partial L_t}{\partial W_h} = \frac{\partial L_t}{\partial y_t} \cdot \frac{\partial y_t}{\partial h_t}\left[\sum_{k=0}^{t}\left(\prod_{i=k+1}^{t} \frac{\partial h_i}{\partial h_{i-1}} \cdot \frac{\partial h_k}{\partial W_h}\right)\right] \tag{8.12}$$

5. Similarly $\frac{\partial L_t}{\partial W_x}$ can be calculated and updated the weight W_x and bias.
6. Repeat until o_t equals y_t.

A simple RNN with single neuron in the hidden layer is considered for understanding the learning limitations. The term $\frac{\partial h_i}{\partial h_{i-1}} \cdot \frac{\partial h_k}{\partial W_h}$ in Equation 8.11 contributes an exponential relation in the weight updation process. The scalar product of $\frac{\partial h_i}{\partial h_{i-1}} \cdot \frac{\partial h_k}{\partial W_h}$ results in scalar change in loss function. If the absolute value of $\frac{\partial h_i}{\partial h_{i-1}} \cdot \frac{\partial h_k}{\partial W_h}$ is less than 1, then the scalar product reaches zero value exponentially. Whereas $\left|\frac{\partial h_i}{\partial h_{i-1}} \cdot \frac{\partial h_k}{\partial W_h}\right|$ is greater than 1, then the product reaches infinity exponentially. In other words, the normalized gradient of the loss function at the output layer grows exponentially either to 1 or 0. While the former is termed as exploding gradient and later as vanishing gradient, both make the learning of correlation between the events difficult to proceed.

In case of vanishing gradient, the contribution of faraway time-steps was lost and hence the learning is implemented with only nearer time-step information. RNN with this limitation is not suitable for long-term-dependent sequences with this structure. If the normalized gradient of loss function rises exponentially, the product will not be a scalar in exploding gradient condition. The learning of RNN with this limitation becomes unstable and the network may get crashed. More the number of layers (k) in hidden state, the normalized scalar values become spectral matrix form which makes the learning complex and difficult. Then, the norm of all the matrix is computed using chain rule to learn gradient issues. Thus, the RNN model is simple and powerful, the training of RNN has vanishing and exploding limitations with many deep layers.

The gradient limitations have been dealt in the following ways:

i. Time-delay neural networks
ii. Leaky units and a hierarchy of different time units
iii. LSTMs, GRUs
iv. Optimization methods
v. Gradient clipping for exploding gradients

vi. Regularization to encourage information flow
vii. L1/L2 penalties

8.5.2.2 LONG SHORT-TERM MEMORY NETWORKS

LSTMs follow an artificial RNN architecture, which has both feed-forward and feedback connections. It can process entire sequence of data at a time, so this method got popularity in the field of DL. It has wide range of applications like connected, unsegment handwriting analysis, speech and video analysis, anomaly detection in network traffic, or intrusion detection systems.

It has one cell and three gates, that is, input, output, and forget gate (as shown in Figure 8.12), where the role of the cell to remember values over arbitrary time interval, whereas gates regulate, the flow of information into the cell or out of the cell or if anything can be dropped out.

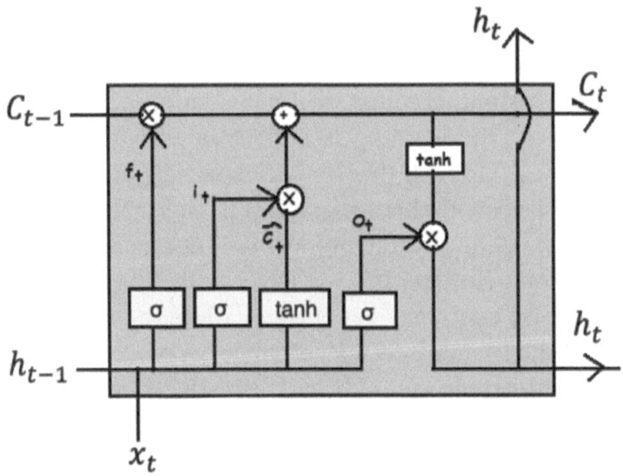

FIGURE 8.12 Long short-term memory networks.

Advantages of LSTMs

1. It can perform processing of dataset, classification from the dataset.
2. Based on times series data, it can make predictions.
3. It has ability to bridge lags of unknown duration between events in time series.

4. It can address the issues of vanishing gradient of traditional RNNs.
5. It is not sensitive to gap length unlike hidden Markov models and other sequence learning techniques.
6. LSTMs are strongly capable of achieving state-of-the-art results in machine translation field.

Disadvantages of LSTMs

1. For fine precision counting time steps to achieve, additional counting methods may be required.
2. Number of gate requirement is more as each memory cell block requires an input and one output gate.
3. It shows constant error as shown in conventional feed-forward architecture while presented entire string at a time as input.
4. Just like other feed-forward networks, LSTMs also have "regency" problem.

Let us consider:
C_t, h_t: hidden layer vectors
x_t: input vector
b_f, b_i, b_c, b_o: bias vector
W_f, W_i, W_c, W_o: parameter matrices
σ, tan h: activation functions

$$f_t = \sigma(W_f \cdot [h_{t-1}, x_t] + b_f) \tag{8.13}$$

$$\breve{i}_t = \sigma(W_i \cdot [h_{t-1}, x_t] + b_i) \tag{8.14}$$

$$o_t = \sigma(W_o \cdot [h_{t-1}, x_t] + b_o) \tag{8.15}$$

$$\breve{C}_t = \tan h(W_c \cdot [h_{t-1}, x_t] + b_c) \tag{8.16}$$

$$C_t = f_t \odot C_{t-1} + i_t \odot \breve{C} \tag{8.17}$$

$$h_t = o_t \odot \tan h(C_t) \tag{8.18}$$

8.5.2.3 GATED RECURRENT UNITS

In RNN, GRUs are gated mechanisms. It has fewer parameters compared to LSTMs with similar architecture with forget gate. GRUs are of two types: (1) fully gated unit and (2) minimal gated unit. It is introduced mainly to prevent vanishing gradient, issue of standard RNN using update and reset gate.

The fully gated unit (as shown in Figure 8.13) has several variations, based on bias combinations and gating done using previous hidden state. The most simplified form of GRU is known as minimal gated unit. Main three types are shown as Type 1, Type 2, and Type 3 in Figures 8.14–8.16, respectively.

Let us consider:

h_t: hidden layer vectors
x_t: input vector
z_t: update gate vector
r_t: reset gate vector
b_z, b_r, b_h: bias vector.
W_z, W_r, W_h: parameter matrices
σ, tan h: activation functions

If $\sigma(x) \in [0,1]$, then alternative activation functions are also possible. The operator $\odot$ denotes the Hadamard product:

$$z_t = \sigma(W_z \cdot [h_{t-1}, x_t] + b_z) \tag{8.19}$$

$$r_t = \sigma(W_r \cdot [h_{t-1}, x_t] + b_r) \tag{8.20}$$

$$h_t = (1 - z_t) \odot h_{t-1} + z_t \odot \tan h(W_h \cdot [r_t \odot h_{t-1}, x_t] + b_h) \tag{8.21}$$

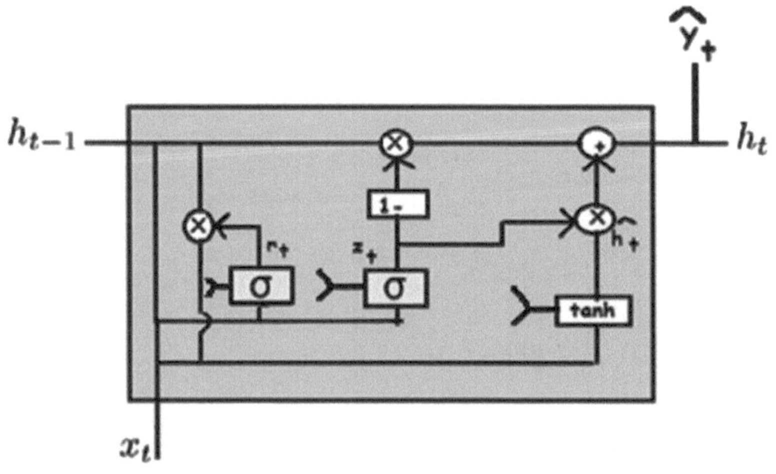

FIGURE 8.13 Fully gated version.

Type 1: Each gate only depends on bias and previous hidden state.

$$z_t = \sigma(W_z h_{t-1} + b_z) \quad (8.22)$$

$$r_t = \sigma(W_r h_{t-1} + b_r) \quad (8.23)$$

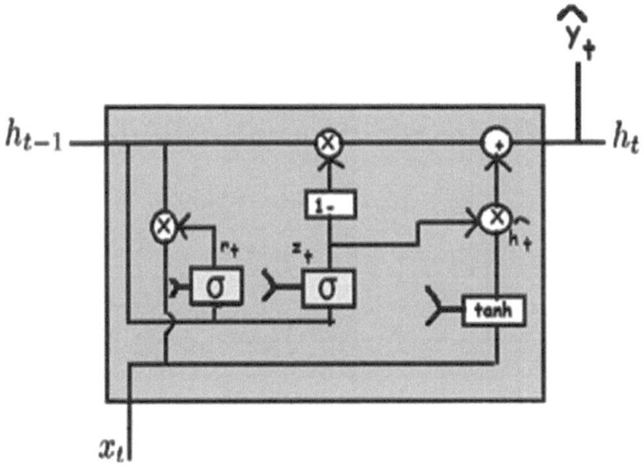

FIGURE 8.14 Type 1.

Type 2: Each gate only depends on previous hidden state without bias.

$$z_t = \sigma(W_z h_{t-1}) \quad (8.24)$$

$$r_t = \sigma(W_r h_{t-1}) \quad (8.25)$$

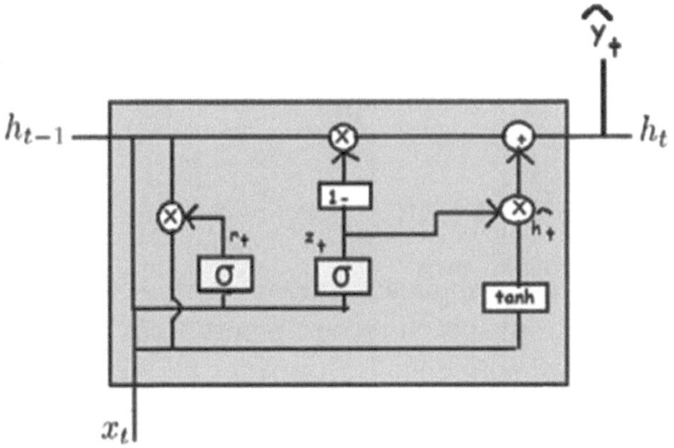

FIGURE 8.15 Type 2.

Type 3: Each gate only depends on bias and not previous hidden state.

$$z_t = \sigma(b_z) \tag{8.26}$$

$$r_t = \sigma(b_r) \tag{8.27}$$

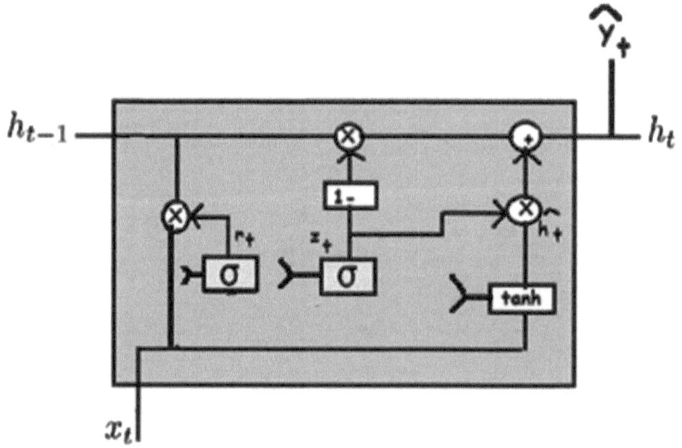

FIGURE 8.16 Type 3.

Minimal Gated Unit

In case of minimal gated unit, the update vector (z_t) and reset vector (r_t) are merged into a forget gate (f_t).

$$f_t = \sigma(W_f \cdot [h_{t-1}, x_t] + b_f) \tag{8.28}$$

$$h_t = f_t \odot h_{t-1} + (1 - f_t) \odot \tan h(W_h \cdot [f_t \odot h_{t-1}, x_t] + b_h) \tag{8.29}$$

where x_t is the input vector; h_t is the output vector; f_t is the forget vector; and W_f, W_h, and b_f, b_h are the parameter matrices and bias vectors.

8.6 APPLICATIONS OF RNN

RNNs are used in many applications ranging from seizure detection to cryptography and from speech recognition to stock market prediction. Very importantly we notice that the most or all these applications are unsegmented, connected, and sometimes temporal in nature and are very much applicable to physiological signal data as they are in the form of one-dimensional (1D) signals and are time-domain data. These data are recorded as samples over time. Therefore, many research works have benefitted from the approach of applying RNNs to physiological data.

Physiological data also known as biomedical signals specifically bioelectrical signal categorizes and characterizes as follows:

- Electroencephalogram (EEG): Electrical signals generated due to electrical activity in human brain.
- Electrooculogram (EOG): Signals generated due to the change of the cornea–retinal potential that exists between the front side and back end of human eye.
- Electrocardiogram (ECG): Signals that are generated from activity of human heart which are consequence of cardiac muscle depolarization and repolarization during heartbeat.
- Electromyogram (EMG): Electrical signals generated by skeletal muscles.

Last two years, that is, 2018 and 2019, it has been seen a lot of research development happened on EEG and also using EEG among the four other modalities or combination of two or more modalities.[38] EEGs are majorly used because EEG-based signal acquisition is cheaper and does not require any special equipment, the acquisition system is portable which makes it feasible for applications such as brain–computer interface; it has high temporal resolution with sampling rates between 250 and 2000 Hz.

Recent past, several experimental and theoretical work on statistical classification techniques had been published. Many of the literatures prove that the DL models outperforming compared to traditional models. In the following section, EEG modality using RNN has been discussed. In the next section, the prominent applications of EEG research such as brain decoding and anomaly detection have been elaborated.

8.7 APPLICATIONS OF EEG SIGNALS AND WITH RNNS

EEG signals are used as a tool to understand the activity or state of human brain. Typically, the voltage potentials are generated due to ionic current within the neurons of human brain. These signals are recorded either by noninvasive method, with electrodes placed on scalp or by invasive method, in which electrodes are implanted through neurosurgery in the brain under the skull.

With the help of several research work, application of EEG signal can be broadly categorized into Brain decoding and Anomaly detection[32] as shown in Figure 8.17.

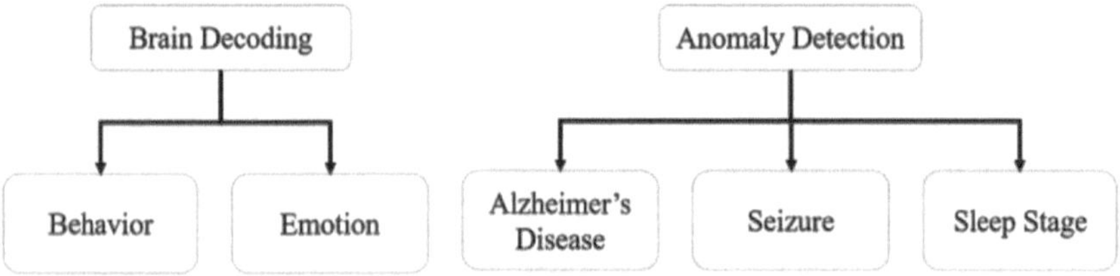

FIGURE 8.17 Classification of areas of EEG signal research.

8.7.1 BRAIN DECODING

Brain decoding is a field of research working toward understanding brain signals recorded from human brain while performing a specific task or interpreting natural human behavior. Characteristic pattern that human brain produces for each activity has been studied and found new solutions to automatically identify or classify these signals based on the pattern they produce. Because of its specific characteristic behavior, these signals are majorly used in the concept of brain–machine interface such as robotics.[5] Furthermore, brain-decoding signals are classified as Behavior signals which relates to the positive, negative, or steady-state events and emotion signals which are used to classify the human's emotions or his/her intension.

Behavior brain decoding involves work relating to evoked potentials such as steady-state evoked potentials (SSEP), P300 signals, and spontaneous signals such as motor cognitive tasks. SSEP signals are natural signals and have characteristic frequency patterns when given a visual stimulus. Similarly, P300 signals are a class of event-related potentials which are generated in the process of decision-making.[4]

Emotion decoding involves work relating to valence arousal classification or specifically identifying person's emotion using EEG signals. Table 8.2 gives an overview of papers for brain decoding using RNN.

TABLE 8.2 Overview of papers for Brain Decoding Using Recurrent Neural Networks.

Medical application	Medical task	DL model	Ref.
MI	Hand movement classification	LSTM	[56]
		LSTM	[47]
	MI recognition	LSTM, DWT-LSTM	[26]
		CNN + GRU	[18]
	MI classification	LSTM, BiLSTM, CNN + RNN	[29]
Evoked potentials	SSVEP classification	CNN + LSTM	[25]
Emotion	Emotion recognition	LSTM	[36]
		CNN + LSTM	[26]
		LSTM	[2]
		RNN	[46]
	Intension recognition	Cascaded CNN and LSTM	[55]
		Cascaded CNN and RNN	[17]
		LSTM	[12]

8.7.2 ANOMALY DETECTION

Anomaly detection is defined as the process of identifying an unusual characteristic in EEG signals. Some of the representative examples include sleep process monitoring, brain damage from head injury, stoke, early detection of epileptic seizure, and many other problems relating to neurological disordering. Some of the work in this domain either work toward classifying the type of anomaly or categorizing anomalies.

One of the subsets of anomaly detection is sleep-stage classification. Sleep disorder can affect the normal life with various neurological diseases. Research includes classifying of nonrapid eye movement sleep (NREM) and rapid eye movement sleep and the stages of NREM sleep such as stages 1, 2, and 3. Table 8.3 gives an overview of papers who have used RNNs for detecting anomalies in EEG signals.

8.8 APPLICATION OF RNN IN SEIZURE CLASSIFICATION

One specific application in which RNNs are used for seizure classification has been discussed: time series multiclass classification of EEG signal using one of the RNN method; LSTM. Two major reasons for selection of LSTM among other techniques; LSTM is chosen because (i) the training of simple RNN architecture is troublesome because of vanishing/exploding gradient issues in deeper layer and LSTM handles this properly and (ii) LSTMs are more accurate on dataset using longer sequence compared to GRU. Hence, in order to demonstrate a classification, LSTM network is written in Python with Keras framework. A multilayer LSTM is trained with various hyperparameters and the best optimal solution is determined based on classification results of EEG for five different classes.

TABLE 8.3 Overview of Papers for Anomaly Detection Using Recurrent Neural Networks.

Medical application	Medical task	DL model	Ref.
Seizure	Seizure detection	Bi LSTM	[48]
		GRU	[14]
		CNN + BiLSTM	[1]
	Seizure classification	LSTM, GRU	[13]
		RNN	[24]
	Seizure prediction	CNN, LSTM	[27]
		LSTM	[49]
Alzheimer	Prediction	BiLSTM	[34]
Parkinson's	Classification	CNN + LSTM	[41]
		LSTM, GRU	[39]
Sleep Stage	Sleep stage classification	CNN + LSTM	[6]
		LSTM	[31]
		BiRNN	[35]
		CNN + GRU	[45]
		Feature + RNN	[8]
		RNN	[40]

In this chapter, focus is on seizure classification.

8.8.1 WHAT IS SEIZURE?

According to the World Health Organization, neural network disorder (commonly known as epilepsy) affects approximately 45–50 million people worldwide (1 million cases in India) in which several cases remain undiagnosed and untreated. Electroencephalogram is one of the electrophysiological methods, records the neurological activity in the brain, and is commonly used to identify network disorder. Depending on seizure frequency, epilepsy syndrome, and site of seizure onset, the sensitivity of EEG may vary quite high. During a seizure, a person may behave abnormally and/or might become unconscious, and so on. These symptoms can be categorized as types of seizures which are clinically practiced. The main different types of seizures are:

1. Febrile seizure
    a. Typical
    b. Atypical
2. Partial seizure
    a. Simple partial or focal seizure
    b. Complex partial seizure
        i. Temporal
        ii. Excited
3. Generalized seizure
    a. Generalized tonic–clonic seizure
    b. Tonic seizure
    c. Clonic seizure
    d. Atonic
    e. Absence seizure
    f. Myoclonic seizure
4. Unclassified

These seizures can be treated with different antiseizure drug (ASDs). However, in clinical setup, it is extremely difficult to categorize seizure based on symptoms because of insufficient information shared by patients and relatives on onset seizure.

In this situation, the diagnosis using EEG signals can be beneficial. The present work aims to classify different types of epilepsy, through electrophysiological investigation using artificial intelligence techniques as error-free diagnosis from EEG is very challenging. These ASDs work well with only generalized epilepsy but accompanied by several side effects. So, it is extremely important to diagnose properly the classes.

On the other hand, ASDs are also administered for partial and pharmacoresistant epilepsy. But they have no effect in controlling the disorder instead they create side effect. The alternate method of treating partial epilepsy or pharmacoresistance epilepsy is to undergo surgery. But this requires proper classification of seizure as well as prediction of the region of network disorder.

From the above discussion, it has been understood that the correctness on interpretation and classification of 1D EEG signal is very important for diagnostic point of view. It is very important to interpret the 1D EEG signals and classify among different activities of brain for various diagnostic purposes. The fully interconnected hidden configuration of RNN makes the model very dominant which enables to correlate between faraway events in the data.

8.8.2 DATASET

Deep-learning method depends on good algorithm and also good dataset for high accuracy. Publicly available dataset known as epileptic seizure recognition dataset from University of California, Irvine (UCI) machine learning repository[3] has been used in this study. It is a preprocessed and restructured dataset used for epileptic seizure detection. The dataset is a multivariate time series data with 179 attributes and 11,500 instances. The data were collected from 500 individuals with 4097 data points for 23.5 s. The data were shuffled and divided for each second resulting in 11,500 samples of data categorized into five groups. Each group is equally sized with 2300 samples which strike out the unbalanced data problem. Figure 8.18 shows the sample for each category and Table 8.4 shows the categories and its description.

8.8.3 IMPLEMENTATION

Figure 8.19 shows the general block diagram of end-to-end learning using DL. Here, the DL method (LSTM) is used as a classifier.

Data are split into two groups with 70/30 split for training and testing, respectively. Eight Thousand and fifty samples were taken for training and 3450 samples for testing. Hence, each sample had a 178-dimensional vector.

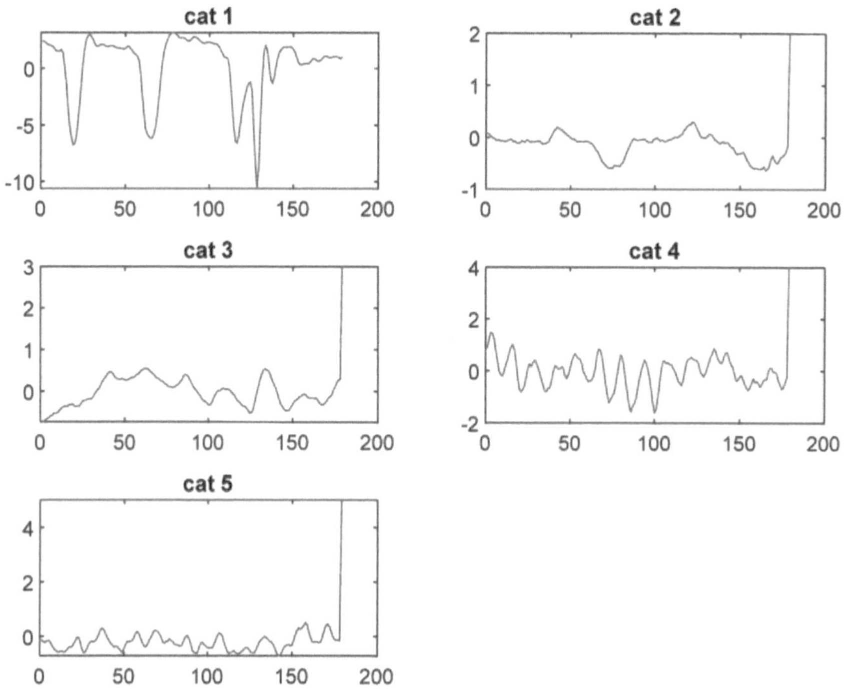

FIGURE 8.18 Sample EEG plot for each category.

TABLE 8.4 Category and Its Description.

Category	Description
1	EEG of network disorder
2	EEG from tumor located area
3	EEG of normal healthy brain
4	EEG signals in eyes closed condition
5	EEG signals in eyes open condition

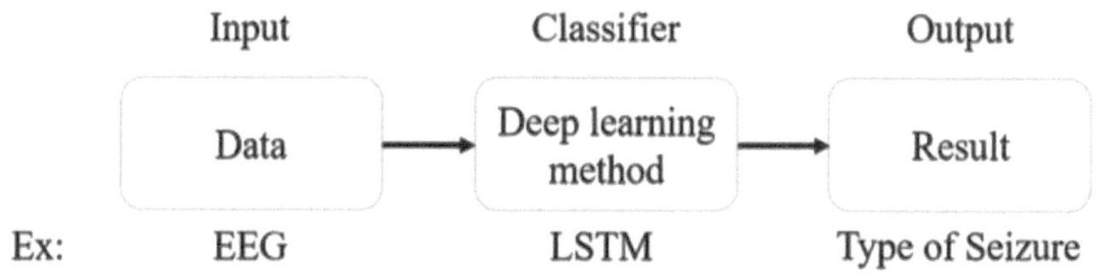

FIGURE 8.19 Deep learning architecture for end-to-end learning.

8.8.4 NETWORK ARCHITECTURE

Unit-wise approach has been adopted in this study (i.e., number of neurons in a single layer), layer wise (number of layers), and model wise (simple RNN, LSTM, GRU). In order to develop an example, two-layered architecture has been considered with 32 LSTM units on first layer and second layer. Figure 8.20 shows the network architecture, where each segment with 178 time steps is given as input followed by an LSTM layer with 32 units which returns a sequence of vectors of dimension 32, followed by another LSTM layer with 32 units that returns a single vector of dimension 32 as described in Kusumika 2019.[22] Finally, a fully connected layer with five layers followed by a softmax function which gives a categorical output. Additionally, layers such as batch normalization and drop-out layers also can be included, which improves the network performance and overfitting.

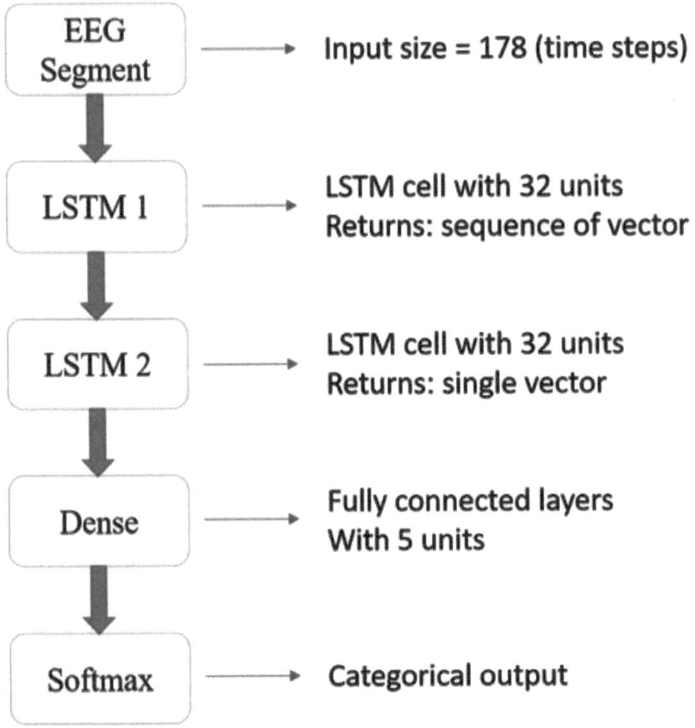

FIGURE 8.20 Network architecture used for seizure detection.

Figure 8.20 shows the network architecture for seizure detection. The first line of code in Figure 8.21 defines a group of linear stack of layers which are appropriate to use when each layer has exactly one input tensor and one output tensor (https://medium.com/@mokarakaya_83469/the-timeline-of-recurrent-neural-networks-and-how-to-use-the-improvements-in-tensorflow-a134e2febe28). Second line and third line add up an LSTM layer to the model with 32 units each. If "return_sequences" are made "True," then it returns a sequence of vectors and if "false" it returns a single vector. This vector is passed through a dense or fully connected layer with softmax activation. BatchNormalization layer is added after LSTM as suggested by Margarit[30] as it improves the training speed and regularizes the model. A dropout layer is added to avoid overfitting. Finally, the last line prints the summary on terminal.

```
model = Sequential()
model.add(LSTM(32, return_sequences=True, input_shape=(178, 1)))
model.add(LSTM(32))
model.add(BatchNormalization())
model.add(Dropout(0.3))
model.add(Dense(5, activation='softmax'))
model.summary()
```

FIGURE 8.21 Code snippet for model creation in Keras python.

8.8.5 *TRAINING*

Once the model is developed, it is compiled and trained to evaluate the performance of the network. Figure 8.22 shows the code snippet of compile, fit, and evaluate. Compile or simply conuring learning process receives three arguments, that is, optimizer, loss function, and a list of matrices. In this example, the type of loss function used is "categorical_crossentropy" which are commonly used in cases where there are many categories or classes where only one case is true. It is noted while using "categorical_crossentropy" function, the labels or target vector needs to be in categorical format and to convert from integers to categorical keras utility "to_categorical" can be used. Next, "Adam optimizer" has been chosen for adaptive stepping. The parameters of optimization are chosen as default values. Finally, "accuracy" is used to evaluate the performance of the model.

```
Layer (type)                 Output Shape              Param #
=================================================================
lstm_1 (LSTM)                (None, 178, 32)           4352
_________________________________________________________________
lstm_2 (LSTM)                (None, 32)                8320
_________________________________________________________________
batch_normalization_1 (Batch (None, 32)                128
_________________________________________________________________
dropout_1 (Dropout)          (None, 32)                0
_________________________________________________________________
dense_1 (Dense)              (None, 5)                 165
=================================================================
Total params: 12,965
Trainable params: 12,901
Non-trainable params: 64
_________________________________________________________________
```

FIGURE 8.22 Output of model summary.

Keras has some handy functions to train the model without writing loops and by passing arguments such as data, labels, batch size, number of epochs, and so on. This function will handle data extraction process, model input, executing gradient steps, and logging metrices. For each iteration, the function picks up a batch of data specified in batch size and train it over, and at the end of each epoch, the validation data are run over the model and the accuracy is calculated. Finally, to evaluate the model performance over the test data, the data and its corresponding targets in categorical format are given. The batch size needs to be same for training and testing as in Figure 8.23.

```
model.compile(loss='categorical_crossentropy',optimizer='adam',
              metrics=['accuracy'])
history = model.fit(X_train, to_categorical(Y_train,num_classes=5),
                    validation_split=0.33, batch_size=128, epochs=100)
score = model.evaluate(X_test, to_categorical(Y_test,num_classes=5),
                       batch_size=128)
```

FIGURE 8.23 Code snippet for compile, fit, and evaluate in Keras.

8.8.6 RESULTS

The model is trained for 100 epochs with a batch size of 128 by splitting 33% of training data for validation after each epoch. Four thousand six hundred and eighty-nine samples are used for training and 2311 samples for validation. Figure 8.24 shows the console output once the training is started for five epochs. The time taken for each step is 3 ms and 12 s for each epoch.

```
Train on 4689 samples, validate on 2311 samples
Epoch 1/100
4689/4689 [==============================] - 12s 3ms/step - loss: 1.5097 - accuracy: 0.3355 - val_loss: 1.4664 - val_accuracy: 0.4111
Epoch 2/100
4689/4689 [==============================] - 11s 2ms/step - loss: 1.2533 - accuracy: 0.4517 - val_loss: 1.3979 - val_accuracy: 0.3977
Epoch 3/100
4689/4689 [==============================] - 11s 2ms/step - loss: 1.1174 - accuracy: 0.5108 - val_loss: 1.2439 - val_accuracy: 0.4760
Epoch 4/100
4689/4689 [==============================] - 11s 2ms/step - loss: 1.0093 - accuracy: 0.5541 - val_loss: 1.0404 - val_accuracy: 0.5634
Epoch 5/100
4689/4689 [==============================] - 11s 2ms/step - loss: 0.9094 - accuracy: 0.5897 - val_loss: 1.0433 - val_accuracy: 0.5500
```

FIGURE 8.24 Console output while training.

Once the training has started, the output of the training process such as record of training loss values and metrics values at every epoch as well as validation accuracy and its loss for every epoch are stored in history variable as shown in Figures 8.25 and 8.26, respectively.

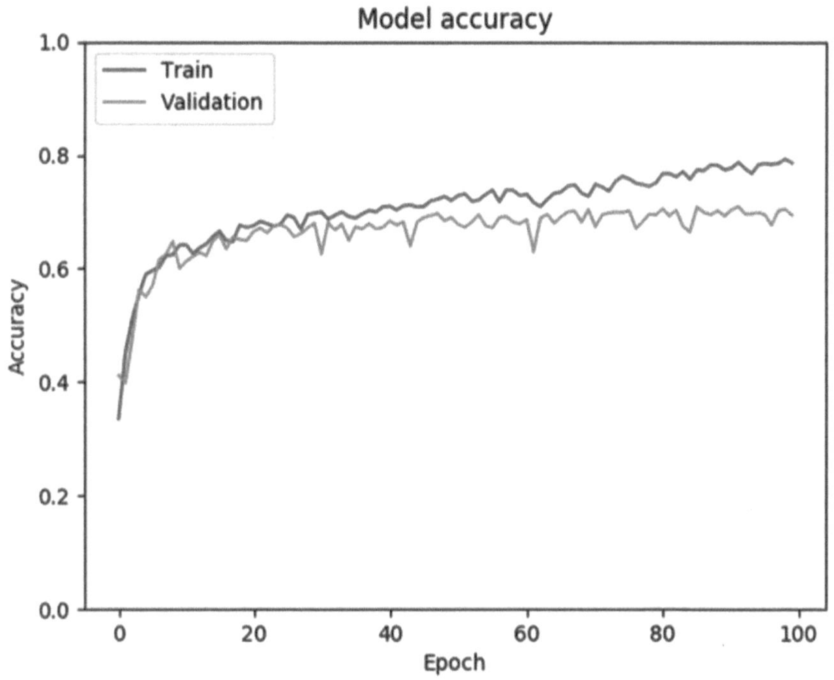

FIGURE 8.25 Plot of model accuracy of training and validation data.

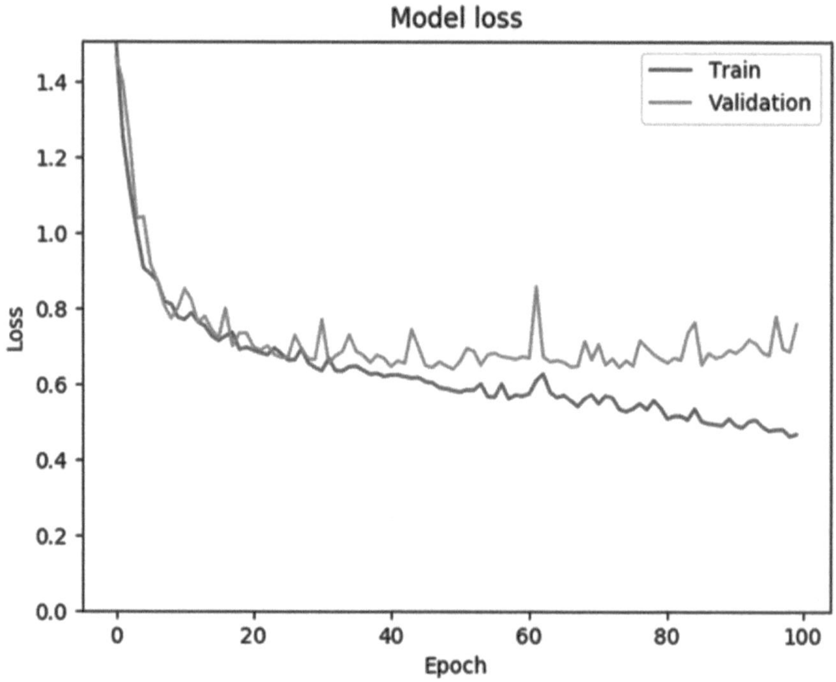

FIGURE 8.26 Plot of model loss on training and validation data.

The model is evaluated after training with the test data which got split at the beginning and not used in training process with its labels has been classified with performance accuracy and loss as 0.696 and 0.77, respectively.

In order to get better performance, various hyperparameters such as optimizer, learning rate, number of epochs, batch size, number of neurons in each layer, and number of layers can be considered. Kusumika[22] has showed the performance variation with number of layers and various hyperparameters. In order to work on any example, it is highly recommended to try using different combinations of abovementioned parameters and hyperparameters.

8.9 SUMMARY

In this chapter, about RNN, it's various architecture, types, and applications have been discussed. For better understanding, one specific example, that is, seizure classification (with preliminary understanding of seizure and its type), has been discussed in details, such that reader can quickly do hands-on experiments.

KEYWORDS

- **RNN**
- **GRU**
- **LSTM**
- **EEG**
- **seizure**
- **multiclass**

REFERENCES

1. Abdel Hameed, A. M.; Daoud, H. G.; Bayoumi, M. Deep Convolutional Bidirectional LSTM Recurrent Neural Network for Epileptic Seizure Detection. In *2018 16th IEEE International New Circuits and Systems Conference (NEWCAS)*. IEEE, 2018, June; pp 139–143.
2. Alhagry, S.; Fahmy, A. A.; El-Khoribi, R. A. Emotion Recognition Based on EEG Using LSTM Recurrent Neural Network. Emotion **2017,** *8* (10), 355–358.
3. Andrzejak, R. G.; Lehnertz, K.; Rieke, C.; Mormann, F.; David, P.; Elger, C. E. Indications of Nonlinear Deterministic and Finite Dimensional Structures in Time Series of Brain Electrical Activity: Dependence on Recording Region and Brain State. *Phys. Rev. E* **2001,** *64*, 061907.
4. Bellary, S. A. S.; Conrad, J. M. Classification of Error Related Potentials Using Convolutional Neural Networks. In *2019 9th International Conference on Cloud Computing, Data Science & Engineering (Confluence)*. IEEE, 2019, January; pp 245–249.
5. Bellary, S. A. S. *Human-Robot Cooperation Using EEG Signals with Self-Learning.* Doctoral Dissertation, The University of North Carolina at Charlotte, 2019.
6. Bresch, E.; Großekathöfer, U.; Garcia-Molina, G. Recurrent Deep Neural Networks for Real-Time Sleep Stage Classification from Single Channel EEG. *Front. Comput. Neurosci.* **2018,** *12*, 85.
7. Hillar, C.; Mehta, R.; Koepsell, K. A Hopfield Recurrent Neural Network Trained on Natural Images Performs State-of-the-Art Image Compression. In *IEEE International Conference on Image Processing (ICIP)*, Paris, 2014; pp 4092–4096.

8. Sun, C.; Fan, J.; Chen, C.; Li, W.; Chen, W. A Two-Stage Neural Network for Sleep Stage Classification Based on Feature Learning, Sequence Learning, and Data Augmentation *IEEE Access* **2019,** *7*, 109386–109397.
9. Aggarwal, C. C. *Neural Networks and Deep Learning—A Textbook*; Springer International Publishing AG, Part of Springer Nature, 2018. https://doi.org/10.1007/978-3-319-94463-0.
10. Cho, K.; van Merriënboer, B.; Gulcehre, C.; Bougares, F.; Chwenk, H.; Bengio, Y. Learning Phrase Representations Using RNN Encoder-Decoder for Statistical Machine Translation, 2014. 10.3115/v1/D14-1179.
11. Rumelhart, D. E.; Hinton, G. E.; Williams, R. J. Learning Internal Representations by Error Propagation. *Parallel Distributed Processing: Explorations in the Microstructure of Cognition, Vol. 1: Foundations*; MIT Press: Cambridge, MA, USA, 1986; pp 318–362.
12. Şen, D.; Sert, M. Continuous Valence Prediction Using Recurrent Neural Networks with Facial Expressions and EEG Signals. In *2018 26th Signal Processing and Communications Applications Conference (SIU),* Izmir, 2018; pp 1–4.
13. Dutta, K. K. 2019, January. Multi-class Time Series Classification of EEG Signals with Recurrent Neural Networks. In *2019 9th International Conference on Cloud Computing, Data Science & Engineering (Confluence)*. IEEE; pp 337–341.
14. Wang, F.; et al. Analysis for Early Seizure Detection System Based on Deep Learning Algorithm. In *2018 IEEE International Conference on Bioinformatics and Biomedicine (BIBM)*, Madrid, Spain, 2018; pp 2382–2389.
15. Gers, F. A.; Schmidhuber, J. A.; Cummins, F. A. Learning to Forget: Continual Prediction with LSTM. *Neural Comput.* **2000,** *12* (10), 2451–2471. https://doi.org/10.1162/089976600300015015.
16. Sutskever, I. *Training Recurrent Neural Networks*. PhD Thesis, University of Toronto, 2013.
17. Miranda-Correa, J. A.; Patras, I. A Multi-task Cascaded Network for Prediction of Affect, Personality, Mood and Social Context Using EEG Signals. In *2018 13th IEEE International Conference on Automatic Face & Gesture Recognition (FG 2018)*, Xi'an, 2018; pp 373–380.
18. Cai, J.; Wei, C.; Tang, X.; Xue, C.; Chang, Q. The Motor Imagination EEG Recognition Combined with Convolution Neural Network and Gated Recurrent Unit. In *2018 37th Chinese Control Conference (CCC)*, Wuhan, 2018; pp 9598–9602.
19. Jaeger, H. The "Echo State" Approach to Analysing and Training Recurrent Neural Networks—With an Erratum Note. *German National Research Center for Information Technology GMD Technical Report*, 2001. Bonn, Germany.
20. Patterson, J.; Gibson, A. *Deep Learning: A Practitioner's Approach*, 1st ed.; O'Reilly Media, Inc., 2017.
21. Dutta, K. K. Multi-class Time Series Classification of EEG Signals with Recurrent Neural Networks. In *IEEE International Conference on Cloud Computing, Data Science and Engineering*, 10–11 Jan'19, Amity University, Noida, 2017.
22. Dutta, K. K.; Kavya, V.; Swamy, S. A. Removal of Muscle Artifacts from EEG Based on Ensemble Empirical Mode Decomposition and classification of Seizure using Machine Learning Techniques. In *IEEE International Conference on Inventive Computing and Informatics (ICICI 2017)* 23–24 November 17, 2019, IEEE Xplore Compliant – Part Number: CFP17L34-ART. ISBN: 978-1-5386-4031-9.

23. Dutta, K. K.; Swamy, S. A. Machine Learning Techniques for Indian Sign Language Recognition. In *IEEE International Conference on Current Trends in Computer, Electrical, Electronics and Communication*; September 8–9, 2017. VVCE, Mysuru, 2017.
24. Naderi, M. A.; Mahdavi-Nasab, H. Analysis and Classification of EEG Signals Using Spectral Analysis and Recurrent Neural Networks. In *2010 17th Iranian Conference of Biomedical Engineering (ICBME), Isfahan*, 2010, pp 1–4.
25. Attia, M.; Hettiarachchi, I.; Hossny, M.; Nahavandi, S. A Time Domain Classification of Steady-State Visual Evoked Potentials Using Deep Recurrent-Convolutional Neural Networks. In *2018 IEEE 15th International Symposium on Biomedical Imaging (ISBI 2018)*, Washington, DC, 2018; pp 766–769.
26. Li, M.; Zhang, M.; Luo, X.; Yang, J. Combined Long Short-Term Memory Based Network Employing Wavelet Coefficients for MI-EEG Recognition. In *2016 IEEE International Conference on Mechatronics and Automation*, Harbin, 2016; pp 1971–1976.
27. Sun, M.; Wang, F.; Min, T.; Zang, T.; Wang, Y. Prediction for High Risk Clinical Symptoms of Epilepsy Based on Deep Learning Algorithm. *IEEE Access* **2018,** *6*, 77596–77605.
28. Goudreau, M. W.; Giles, C. L.; Chakradhar, S. T.; Chen, D. First-Order Versus Second-Order Single-Layer Recurrent Neural Networks. *IEEE Trans. Neural Netw.* **1994,** *5* (3), 511–513.
29. Ma, X.; Qiu, S.; Du, C.; Xing, J.; He, H. Improving EEG-Based Motor Imagery Classification via Spatial and Temporal Recurrent Neural Networks. In *2018 40th Annual International Conference of the IEEE Engineering in Medicine and Biology Society* (EMBC). IEEE, 2018, July; pp 1903–1906.
30. Margarit, H.; Subramaniam, R. A Batch-Normalized Recurrent Network for Sentiment Classification. *Adv. Neural Inf. Process. Syst.* **2016,** 2–8.
31. Michielli, N.; Acharya, U. R.; Molinari, F. Cascaded LSTM Recurrent Neural Network for Automated Sleep Stage Classification Using Single-Channel EEG Signals. *Comput. Biol. Med.* **2019,** *106*, 71–81.
32. Min, S.; Lee, B.; Yoon, S. Deep Learning in Bioinformatics. *Brief. Bioinf.* **2017**. doi:10.1093/bib/bbw068.
33. Musau, P.; Johnson, T. T. Verification of Continuous Time Recurrent Neural Networks (Benchmark Proposal). In *ARCH@ADHS*, 2018; pp 196–207.
34. Pan, Q.; Wang, S.; Zhang, J. Prediction of Alzheimer's Disease Based on Bidirectional LSTM. *J. Phys.: Conf. Ser.* **2019,** *1187* (5), 052030.
35. Phan, H.; Andreotti, F.; Cooray, N.; Chén, O. Y.; De Vos, M. Automatic Sleep Stage Classification Using Single-Channel EEG: Learning Sequential Features with Attention-Based Recurrent Neural Networks. In *2018 40th Annual International Conference of the IEEE Engineering in Medicine and Biology Society (EMBC)*. IEEE, 2018, July; pp 1452–1455.
36. Jeevan, R. K.; Venu Madhava Rao, S. P.; Shiva Kumar, P.; Srivikas, M. EEG-Based Emotion Recognition Using LSTM-RNN Machine Learning Algorithm. In *2019 1st International Conference on Innovations in Information and Communication Technology (ICIICT)*, Chennai, India, 2019; pp 1–4.

37. Socher, R.; Chiung-Yu Lin, C.; Ng, A. Y.; Manning, C. D. Parsing Natural Scenes and Natural Language with Recursive Neural Networks. In *Proceedings of the 28th International Conference on International Conference on Machine Learning (ICML'11)*. Omni Press: Madison, WI, USA, 2011; pp 129–136.
38. Rim, B.; Sung, N. J.; Min, S.; Hong, M. Deep Learning in Physiological Signal Data: A Survey. *Sensors* **2020,** *20* (4), 969.
39. Ruffini, G.; Ibañez, D.; Castellano, M.; Dubreuil-Vall, L.; Soria-Frisch, A.; Postuma, R.; Gagnon, J. F.; Montplaisir, J. Deep Learning with EEG Spectrograms in Rapid Eye Movement Behavior Disorder. *Front. Neurol.* **2019,** *10.*
40. Hartmann, S.; Baumert, M. Automatic A-Phase Detection of Cyclic Alternating Patterns in Sleep Using Dynamic Temporal Information. *IEEE Trans. Neural Syst. Rehab. Eng.* **2019,** *27* (9), 1695–1703.
41. Lee, S.; Hussein, R.; McKeown, M. J. A Deep Convolutional-Recurrent Neural Network Architecture for Parkinson's Disease EEG Classification. In *2019 IEEE Global Conference on Signal and Information Processing (GlobalSIP)*, Ottawa, ON, Canada, 2019; pp 1–4.
42. Li, S.; Li, W.; Cook, C.; Zhu, C.; Gao, Y. Independently Recurrent Neural Network (IndRNN): Building A Longer and Deeper RNN. In *IEEE/CVF Conference on Computer Vision and Pattern Recognition*, Salt Lake City, UT, 2018; pp 5457–5466.
43. Schuster, M.; Paliwal, K. Bidirectional Recurrent Neural Networks. *IEEE Trans. Signal Process.* **1997,** *45* (11), 2673–2681. doi:10.1109/78.650093.
44. Hochreiter, S.; Schmidhuber, J. Long Short Term Memory. *Neural Comput.* **1997,** *9* (8), 1735–1782.
45. Sm, I. N.; Zhu, X.; Chen, Y.; Chen, W. Sleep Stage Classification Based on EEG, EOG, and CNN-GRU Deep Learning Model. In *2019 IEEE 10th International Conference on Awareness Science and Technology (iCAST)*. IEEE, 2019; pp 1–7.
46. Zhang, T.; Zheng, W.; Cui, Z.; Zong, Y.; Li, Y. Spatial–Temporal Recurrent Neural Network for Emotion Recognition. *IEEE Trans. Cybern.* **2019,** *49* (3), 839–847.
47. Tayeb, Z.; Fedjaev, J.; Ghaboosi, N.; Richter, C.; Everding, L.; Qu, X.; Wu, Y.; Cheng, G.; Conradt, J. Validating Deep Neural Networks for Online Decoding of Motor Imagery Movements from EEG Signals. *Sensors* **2019,** *19* (1), 210.
48. Thara, D. K.; Prema Sudha, B. G.; Xiong, F. Epileptic Seizure Detection and Prediction Using Stacked Bidirectional Long Short Term Memory. *Pattern Recogn. Lett.* **2019,** *128*, 529–535.
49. Tsiouris, K. M.; Pezoulas, V. C.; Zervakis, M.; Konitsiotis, S.; Koutsouris, D. D.; Fotiadis, D. I. A Long Short-Term Memory Deep Learning Network for the Prediction of Epileptic Seizures Using EEG Signals. *Comput. Biol. Med.* **2018,** *99*, 24–37.
50. Tutschku, K. *Recurrent Multilayer Perceptrons for Identification and Control: The Road to Applications*. Research Report. University of Würzburg, 1995.
51. Li, X.; Song, D.; Zhang, P.; Yu, G.; Hou, Y.; Hu, B. Emotion Recognition from Multichannel EEG Data through Convolutional Recurrent Neural Network. In *2016 IEEE International Conference on Bioinformatics and Biomedicine (BIBM)*, Shenzhen, 2016, pp 352–359.
52. Glorot, X.; Bordes, A.; Bengio, Y. Deep Sparse Rectifier Neural Networks. In *Proceedings of the Fourteenth International Conference on Artificial Intelligence and Statistics, PMLR* **2011,** *15*, 315-323.

53. Bengio, Y.; Simard, P.; Frasconi, P. Learning Long-term Dependencies with Gradient Descent Is Difficult. In *IEEE Trans. Neural Netw.* **1994,** *5* (2), 157–166.
54. Yamashita, Y.; Tani, J. Emergence of Functional Hierarchy in a Multiple Timescale Neural Network Model: A Humanoid Robot Experiment. *PLoS Comput. Biol.* **2008,** *4* (11), e10002201–e10002218. https://doi.org/10.1371/journal.pcbi.1000220
55. Zhang, D.; Yao, L.; Zhang, X.; Wang, S.; Chen, W.; Boots, R.; Benatallah, B. Cascade and Parallel Convolutional Recurrent Neural Networks on EEG-Based Intention Recognition for Brain Computer Interface. In *Thirty-Second AAAI Conference on Artificial Intelligence*, 2018.
56. Zhang, G.; et al. Classification of Hand Movements from EEG Using a Deep Attention-Based LSTM Network. *IEEE Sens. J.* **2020,** *20* (6), 3113–3122.

CHAPTER 9

Brain Tumor Classification Using Convolutional Neural Network

M. JAYASHREE[1*], POORNIMA SRIDHARAN[2], V. MEGALA[3], and R. K. PONGIANNAN[4]

[1]*Anna University Regional Campus, Coimbatore, Tamil Nadu, India*

[2]*Anna University, CEG Campus, Chennai, Tamil Nadu, India*

[3]*SRM Institute of Science and Technology, Ramapuram Campus, Chennai, Tamil Nadu, India*

[4]*SRM Institute of Science and Technology, Katankulathur, Chennai, Tamil Nadu, India*

[*]*Corresponding author. E-mail: shrihema17@gmail.com*

ABSTRACT

Deep learning (DL) network is prioritized for accuracy at higher levels and inherent automated feature extraction along with large amount of labeled data and computing power. In medical application, deep learning is used to detect cancer cells automatically. Deep learning consists of more than 150 hidden layers whereas neural network consists 3 hidden layers. Deep Convolutional Neural Network (DCNN) has achieved a great success in computer vision. Stirring by the structure of visual cortex, CNNs embedded with multiple hidden convolutional layers between the input and output layers have a capability of extracting higher level representative features and non-linear properties. Combination of DL with CNN has excellent results on medical field including classification of skin cancer,

Deep Learning in Visual Computing and Signal Processing. Krishna Kant Singh, Vibhav Kumar Sachan, Akansha Singh & Sanjeevikumar Padmanaban (Eds.)

diabetic retinopathy detection and brain timer segmentation. DCNN is used to classify images of image net with help of an Alexnet model. By convolving small filters with the input patterns, features extraction is done followed by selection of the most distinguishing features and then start to train the classification network. In this paper, different classifiers such as Alex net, Google Net and ResNet are chosen for their error rate architectures. The malignant tumor images that occur in the spinal cord and brain at different stages are the input to the classification system. The main features extracted during pre-processing fed as input to the classifier networks which has been already trained. To evaluate the extracted characteristics, the performances of the classifiers are validated based on architecture, repetition, time consumption and accuracy with respect to number of iterations.

9.1 OVERVIEW OF DEEP LEARNING (DL)

Artificial intelligence (AI) is a branch of computer science, entail in creating intelligent machines which can behave like a human. It includes learning, understanding, reasoning, problem solving, decisions making, etc. like human being. AI includes multidisciplinary sciences like computer vision, neuroscience, biology, psychology, sociology, modeling, and philosophy.

Machine learning (ML) is the subset of AI as shown in Figure 9.1. ML achieves AI with specific algorithm and improves automatically through trained data.

DL is the subset within ML. DL is constructed using artificial neural network (ANN) algorithm with more number of hidden layers (nonlinear), big data, and powerful computational resources, transfer higher level information to the lower layers, and vice versa in order to develop more complex relationships.

The layered neuron connection of the brain is the concept utilized by DL to design the structure, algorithm, learning, and training. Increasing number of such layers increases the depth of model and hence improves the performance of the whole DL network. Application of DL is tremendous; few examples are bioinformatics, medical image processing, speech recognition, computer vision, social network filtering, fraud detection, and customer relationship management.

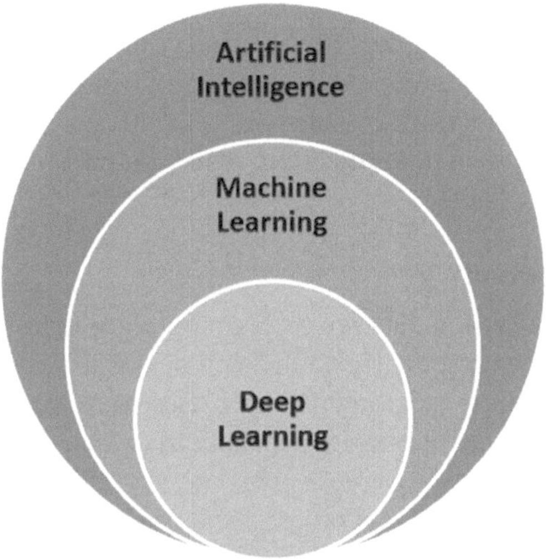

FIGURE 9.1 Relationship between DL, ML, and AI.[6]

9.2 GENERAL ARCHITECTURE OF DL

DL has several algorithms and it is not restricted for particular application. Some algorithms are best fit to specific task. Following are the DL algorithms [Simplelearn link]:

i. Multilayer perceptron neural network (MLPNN)
ii. Back propagation (BP)
iii. Convolutional neural network (CNN)
iv. Recurrent neural network (RNN)
v. Long short-term memory
vi. Generative adversarial network (GAN)
vii. Restricted Boltzmann machine (RBM)
viii. Deep belief network (DBN)

Among the abovementioned algorithm, CNN is considered to best fit for DL concept. Many research works have been carried out using CNN.

i. Multilayer perceptron neural network (MLPNN)

MLPP is constructed using two hidden layers with more than one perceptron and feed forward supervised learning algorithm. Root mean square

(RMSE) is used for error detection. The main advantage of MLPNN is classification of nonlinear data to solve any complex issues.

MLNPP is best suited for the instances which need supervised learning and parallel distributed processing. Some of its applications are image/speech recognition and verification, data classification, e-commerce, machine translation, etc.

ii. Back propagation (BP)

BP is constructed using supervised learning algorithm; optimized using gradient descent with the weights updated backwards to input. The error is sent back to adjust the weights and biases to reduce upcoming error values. The layers in BP operate sequentially which adopt chain rule of derivatives in calculus.

It is best fit for error prone projects and used to train deep ANN. Few applications are image/speech recognition, data mining, and in places of quick derivative requirement.

iii. Convolutional Neural Network (CNN)

CNN is constructed using perceptrons of supervised learning with multi-layer feed forward network to analyze data. Image classification property of CNN leads to visual data applications. Deep CNN architecture AlexNet plays a major role to accelerate research in DL for past several years. CNN finds its application in graph convolution network, image recognition, and text analysis. It includes significant weights and biases to perform differentiation and kernels with minimum preprocessing. Features with less significance are processed in first convolution layer; more significant features are processed in next convolution layer and create a network with sophisticated analysis.

CNN has fewer training parameters to be trained easily and scalable when compared with BPs. Applications of CNN includes image/video processing, recognition, and classification, pattern processing, medical image analysis, natural language processing.

iv. Recurrent Neural Network (RNN)

The RNN is designed to predict the next expected scenario using datasets sequential attribute and patterns. Stochastic gradient descent (SGD) is used in RNN to train the network in addition to BP. Inputs and outputs of RNN are independent such that the hidden layer derives features from previous steps with similar weights as well as bias repeatedly for predictions. All the layers are combined to form single RNN. Information in RNN is processed through feed forward, that is, from input to output and feedback loop using BP.

RNN can process sequential and temporal data. Image/speech recognition, video classification, machine translation-search prediction, natural language processing, and sentiment classification are some of the applications of RNN.

v. Long-Short Term Memory (LSTM)

LSTM uses deep RNN to train instead of gradients which results in unstable condition. Patterns can be recalled or deleted using stored data for long extended period of time. RNN can add, remove, or modify data as per the requirement.

Based on time series data and large stacked network, RNN is suited for classification and prediction efficiently. Sentiment analysis, stock market appreciation, language translation/modeling, and image/video captioning are few applications of LSTM.

vi. Generative Adversarial Network (GAN)

GAN is an unsupervised learning model with robustness. GAN automatically discovers new regularities and patterns and generates a new data for the existing input data. It can also mimic any dataset. GAN has two nets pitting one against the other. Generator and discriminator are the two submodels. The former generates new data while the latter distinguishes between real and false data.

GAN generates new images from given set of images, captures given data, creates a good quality of data, and modifies the data. Speech processing, cyber security, health diagnosis, and natural language processing are the applications of GAN.

vii. Restricted Boltzmann Machine (RBM)

RBM is a probabilistic and graphical model or a stochastic type neural network. It consists of more filtering concept with constrained connections between layers in order to obtain efficient learning. RBM can be replaced by varying auto encoders.

RBM has a visible layer, a hidden layer, and bias unit connected to all the units. Above all there is no connection between the nodes within the group. RBM offers design flexibility and is used for statistical model, classification, regression, and generative models.

viii. Deep Belief Network (DBN)

A DBN is an unsupervised and generative learning model which is also probabilistic. It is a combination of undirected RBM on top layer and is directed in the lower layer for fine tuning. DBN has multilayer of hidden units, processes only one layer sequentially from its base.

DBN is used for energy-based learning, such as image/video recognition and classify satellite image.

9.3 CNN ARCHITECTURE

Basic CNN architecture mimics the pattern of neurons within the human brain. The neurons in CNN are designed as 3D structure and every set of neurons evaluate a small section of the image. In other words, each group of neurons specializes in identifying one part of the image. The CNN has following components (Charu, 2018).

i. Convolutional layer

The convolutional layer is the fundamental building block of CNN because of the heavy computational performance. It creates a feature map. A filter or kernel scans a few pixels of the whole image and predicts the class origin of each feature map.

ii. Pooling layer

It performs down sampling of the information generated by the convolutional layer per feature map and conserves the vital information of inputs. The two types of pooling are:

Max Pooling: It returns the maximum value from the convoluted portion of the image kernel obtained as 4 in pooling output (highlighted in blue) of Figure 9.2 and acts as a noise suppressant.

Average Pooling: It returns the average of all the convoluted output values and performs dimensionality reduction as a noise suppressing mechanism.

iii. Fully connected (FC) layer

FC input layer fetches the previous layer output and creates a single vector which is supplied to next layer. It applies weights to predict an accurate label for the input generated by feature analysis. FC output layer generates the concluding probabilities to identify the class of image.

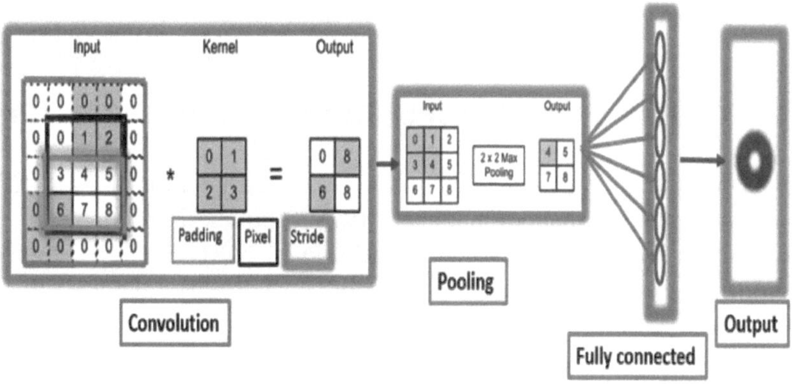

FIGURE 9.2 Basic architecture of CNN (https://d2l.ai/chapter_convolutional-neural-networks/conv-layer.html).

iv. Strides

The granularity level can be reduced using strides along the layers of CNN. Stride is the number of pixels shift over the image matrix. The stride of size 1, 2, and 4 are chosen according to the requirement. Larger stride is preferred to reduce overfitting and compute complex features of large spatial data.

v. Padding

Convolution may reduce spatial size and information along the image edges may lose. Padding adds zero pixel values to maintain the size of convolution features. Half and full padding have been done according to requirement to maintain actual space features without loss.

vi. Training

BP algorithm is used to train convolution, ReLU, and pooling layers of CNN. Errors travel back toward input layer through the aggregated product of loss derivative and filter element in each cell of convolution matrix, straight forward ReLU layers, and maximized pools of CNN.

vii. Data augmentation

Data augmentation is preferred to reduce overfitting in the network performance. New image samples are fed by transforming original image as translated, rotated, patch extracted, and reflected as mirror generated using principal component analysis (PCA) method. The computation of principal components using covariance matrix with Gaussian noise synthesized new samples and helps to reduce error rate by 1%.

9.4 POPULAR ARCHITECTURE OF DL FOR TUMOR DETECTION

The primary tumors (brain originated) or secondary tumors are identified using magnetic resonance imaging (MRI) technique in order to get the information for further treatment. The role of MRI images of brain is prominent to provide the information about the normal or abnormalities in the medical researches. The segmentation of brain images is tedious such that it needs highly professional knowledge as the shape, location, and modality of tumor will not be similar for patients. Clinical experienced experts seek computerized techniques for automatic analysis on tumor segmentation. Identification of tumor using MRI consumes more time for large data. With the specific features of SVM and ANN, the medical researchers used it to affirm the presence of tumors in the last few decades. Researchers classified BRATS images by increasing the number of layers, data augmentation, and preprocessing strategies, normalized preprocessing steps and multiview/multiscale architectures to obtain good accuracy and classification performance.[13]

The successful DL contribution (2D and 3D CNNs) attributes image classification, prediction, and segmentation from the perspective of application, methodology, framework, and algorithms. The tumor images are processed to identify the cell regions, removal of noise, etc in the first stage (BING method), segmented through various techniques (Voronoi diagram, watershed transform, binarization) and classified using deep neural networks (DNN) combined with accuracy improvement techniques (ELM-LRF). The DL with its promising architecture along with feature extraction techniques enrolled an important study in bioinformatics, medical image analysis, etc. The development of DL structures in tumor detector is listed in Table 9.1 as follows in terms of architecture, technique, and tumor data.

TABLE 9.1 DL Architectures for Seizure Classification.

S. no	Methodology	Tumor	Image segmentation	Image classification
1.	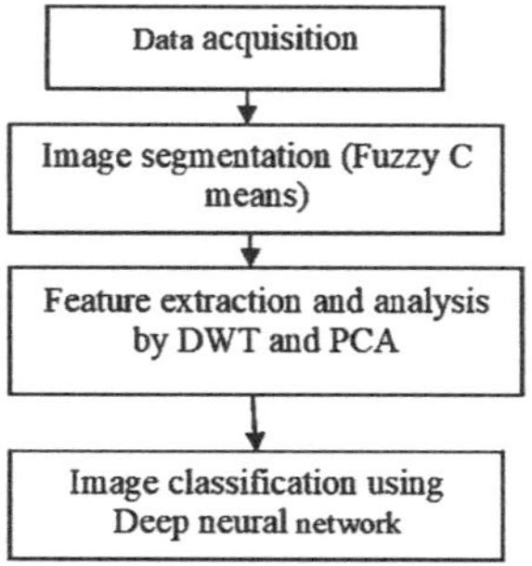 Mohsen et al., 2017	Glioblastoma, metastatic bronchogenic carcinoma tumors and sarcoma Dataset: 66 brain MRIs splitted as 22 nontumor and 44 tumor images (256 × 256) collected from Harvard Medical School website.	Images containing gray matter (GM), cerebrospinal fluid (CSF), white matter (WM), and the skull from the tumor tissues are obtained using Fuzzy clustering technique.	DNN architecture with seven hidden layer structure outperforms the classification among other machine learning classifiers such as KNN; linear discriminant analysis (LDA), and SVM are used for classification. Performance metrics: Classification rate, recall, precision, F-Measure, AUC (ROC).
2.	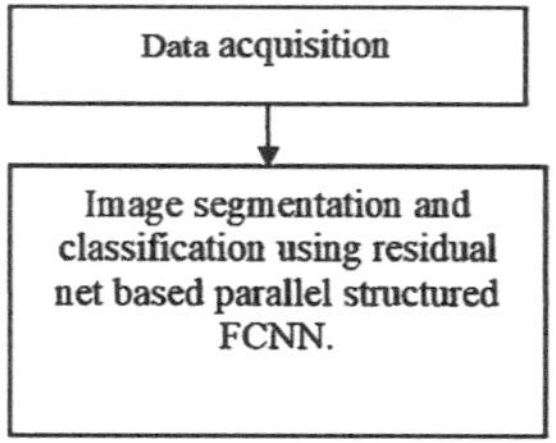 [9]	Gliomas Dataset: High grade gliomas (HGG) dataset includes 220 patients' MRI images and low grade gliomas dataset includes 54 patients' MRI images, each consisting of T1, T2, T1c, and FLAIR types (155 × 240 × 240 × 4). Data is obtained from BRATS 2013 website.	The U-Net residual structures of FCNN combine local and global information through concatenate structure effectively performs semantic segmentation.	Fully connected CNN with residual structures along with hierarchical dice loss for LGG dataset outperformed other classifiers such as VGG+ skip connection, U-Net, ResNet50+ skip connection, and Residual U-Net with boot-strapping loss. Batch normalized layers located between convolutional and activation layers improve training. Performance metrics: Mean of intersection of union (mIoU) and dice score coefficient (DSC), precision, and recall.

TABLE 9.1 *(Continued)*

S. no	Methodology	Tumor	Image segmentation	Image classification
3.	Data acquisition → Image segmentation and classification using residual net based parallel structured FCNN. [10]	Glioma, meningioma, and pituitary tumor. Dataset: 3064 brain images for grading the tumors (128 × 128, 64 × 64, 32 × 32 of three classes) from https://figshare.com/ articles/brain_tumor_dataset/1512427/5.	Image segmentation and classification has been done by CNN and provide confusion matrices for each dataset.	The CNN structure has 18 layers and differentiated graded tumors effectively as DWT, GLCM, texture and shape, intensity histogram GLCM, and bow feature sets. Performance metrics: Accuracy, sensitivity, specificity, and precision.
4.	Data acquisition → Image segmentation using NS-EMFSE approach → Image classification using CNN with classifiers SVM and KNN [11]	80 benign and 80 malignant tumors Dataset: Cancer Genome Atlas Glioblastoma Multiforme (TCGA-GBM) data from The Cancer Imaging Archives (TCIA)	Segmentation of images was done using neutrosophic set—expert maximum fuzzy-sure entropy (NS-EMFSE) approach. The T1 images are converted to gray, filtered with adaptive Weiner filter (AWF) and then converted into binary with neutrosophic algorithm.	The segmented features of MRI images in the classification stage of fivefold AlexNet-based CNN are done using SVM and KNN classifiers. Performance metrics: Sensitivity, precision, accuracy and, Youden index.
5.	Data Acquisition → Image classification using CNN [12]	Tumor and nontumor images. Dataset: Radiopaedia and Brain Tumor Image Segmentation Benchmark (BRATS) 2015	The convolution layer of CNN has small kernels which divides the input image into smaller ones.	The loss function has been chosen as gradient descent to improve the accuracy. The classified output of CNN is found to be superior over SVM classifier.

TABLE 9.1 (Continued)

S. no	Methodology	Tumor	Image segmentation	Image classification
				Performance metrics: Training accuracy, validation accuracy, and validation loss.
6.	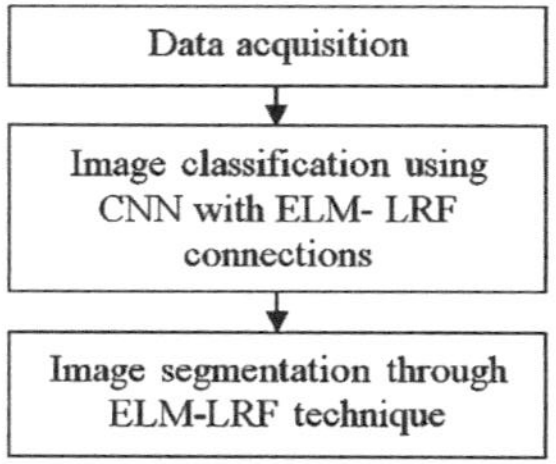 [14]	Benign and Malignant tumors Dataset: 16 cranial MR images are taken from https://medlineplus.gov/mriscans.html and processed as 256 × 256 sized data for classification.	Watershed with morphology process technique is used for image segmentation. It provides the gray density with its defined borders by differentiating the first derivative of pixel changes. Morphological operators help to fix the pixel pointers whose location and quantity determines image segmentation.	The CNN developed for image classification has extreme learning machine (ELM) in the single hidden layer to get zero error in weight updation. The local receptive field (LRF) feature updated the weight vector connecting pooling and output layer by least square method. The combined ELM-LRF classification has been compared with Gabor wavelet, statistical features, and DNN with AlexNet. The developed CNN model has proven superior over SVM, KNN, and NN nets. Performance metrics: Accuracy, sensitivity, and specificity.
7.	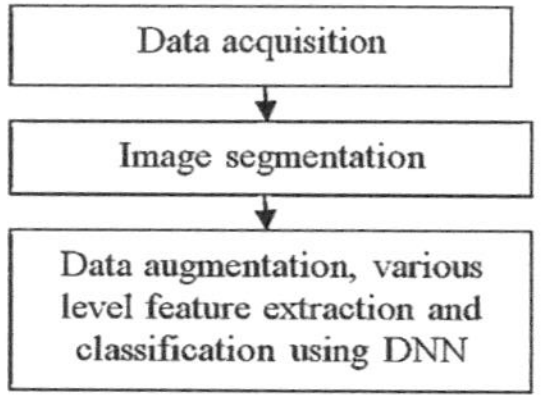 [15]	Meningioma, glioma, pituitary Dataset: 3064 MRI images from 233 persons, out of which 1426 are glioma, 708 meningioma, and 930 pituitary tumors from https://figshare.com.	A free web-based software in various programming languages (http://biostatapps.inonu.edu.tr/BTSY/) provides segmentation automatically for the given dataset.	The free web-based software provides CNN-based classification automatically for the given dataset.

TABLE 9.1 *(Continued)*

S. no	Methodology	Tumor	Image segmentation	Image classification
	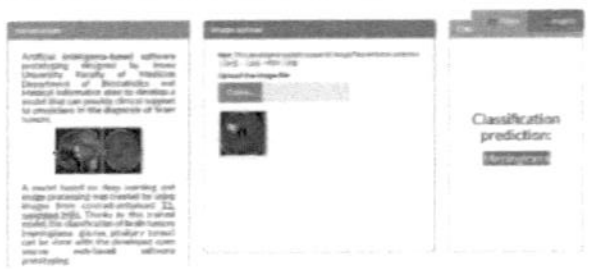			Performance metrics: Matthew's Correlation Coefficient (MCC), F-score, G-mean, accuracy, specificity, sensitivity, and precision.
8.	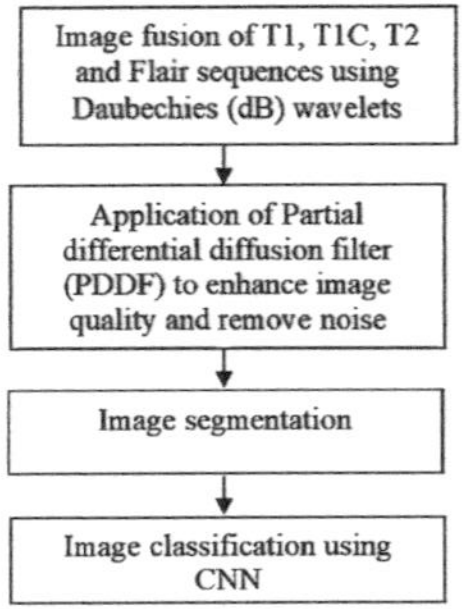 [16]	Glioma Dataset: BRATS 2012, 2013, 2015, 2018, 2013 leader board dataset are used for validating the methodology.	A combination of structural and texture informative images are fed to segmentation process. Global thresholding segments the fused image segments.	The CNN with 23 layers has 6 convolution layers. Each batch normalization layer has been supplied with 8, 16, 32, 64, and 128 channel input segments. The proposed fused input MRI segmentation helps CNN to perform better classification. Performance metrics: Validation error and loss, accuracy, false positive and negative rate.
9.	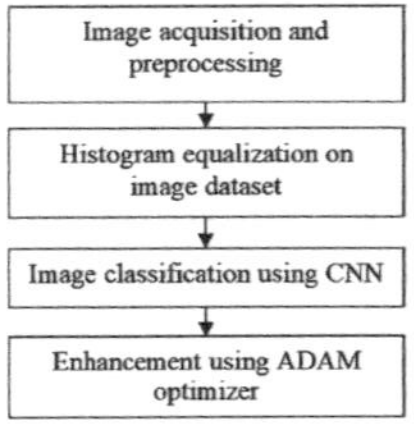 [17]	Meningioma, glioma, and pituitary. Dataset: 3064 images (512 × 512) from 233 cancer patients as meningioma (708 images), pituitary tumor (930 images) and glioma (1426 images).	The original image (512 × 512) has resized to (112 × 112) and filtered to (5 × 5) using Gaussian filter. The filtered data got enhanced the intensity distribution of the filtered image further using histogram equalization.	The three convolution layers of CNN have 32, 64, and 128 filters. The CNN output has further optimized using Adam (AdaGrad-Adaptive Gradient Algorithm and RMSProp-Root Mean Square Propagation) optimizer. The optimized classified output outperforms content-based image retrieval (CBIR) methods.

TABLE 9.1 *(Continued)*

S. no	Methodology	Tumor	Image segmentation	Image classification
				Performance metrics: Support and receiver operating characteristic curve. Confusion matrix, f1-score, precision, and recall.
10.	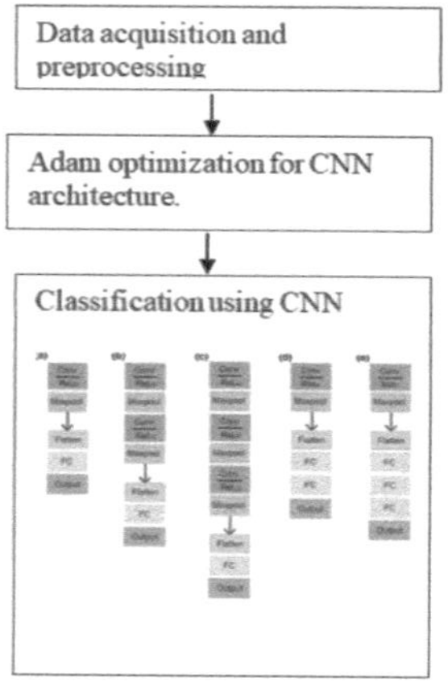 [18]	Meningioma, glioma, and pituitary. Dataset: 3064 MRI images from the website of figshare Cheng (Brain Tumor Dataset, 2017)	Image segmentation and classification has been done by CNN.	The architecture (b) of CNN with 32 filters has better accuracy with low validation loss than others. If the accuracy value of training and validation data increases during consequent epochs of training, then the architecture is said to be good fit. The classifier model is assumed to be as overfit, as the accuracy of validation set tends to decrease and that of training set increases. Performance metrics: Training loss and accuracy, overfitting, validation loss, and accuracy.

9.5 DL CLASSIFIERS FOR SEIZURE SIGNALS

A combination of padded convolutional layer with data augmented and pooling layer together form the ith layer of a CNN. For complex seizure images, layers of whole structure got increased for extracting various level details which also has the limitation of more computational power. The following architectures of CNN are proposed for seizure classification.

- LeNet
- AlexNet
- VGGNet
- GoogLeNet
- ResNet

While developing a model, it takes more time to build a model from the scratch. To reduce the time consumption we can use the pretrained model which has been created for other problem. It reduced the task of creating a new model from the scratch. We can directly use the weights and architecture obtained by training the large datasets in the pretrained model. ImageNet data has been used widely to build various architectures. Fine-tuning helps us to modify the preexisting model. There is no need to modify the weights. To find tune the model, we can use the feature extraction. We can remove the last layer/output layer and then use the entire network.

Freeze other layer while training some layers of the new dataset. The frozen layer keeps the weights of the initial layer while retrain the higher layers.

i. LeNet

LeNet is the first CNN introduced by Lecun et al. in 1998 with 60 thousand parameters. It is mainly used for OCR and character recognition documents. It can be executed in using central processing unit (CPU) instead of graphical processing unit (GPU).

LeNet-5 is constructed using 7 layers (input layer not taken in account) as shown in Figure 9.3. The LeNet-5 consists of two sets of convolutional and average pooling layers, followed by a flattening convolutional layer and two fully connected layers. It uses tanh activation function for all layers, except the last FC layer uses the softmax function.

ii. AlexNet

In 2012, Alex Krizhevsky introduced AlexNet with 60 million parameters. It has won the ImageNet Large Scale Visual Recognition Challenge (ILSVRC) 2012 by a phenomenally large margin. AlexNet was concurrently trained on two Nvidia GeForce GTX 580 GPUs for 6 days as the network seems to be split into two pipelines. It is an essential CNN architecture with first five layers as convolutional layer and next three layers as FC as shown in Figure 9.4. In between the two types of layers, pooling and activation layers are present. It has 15.3% as top five error rate in ILSVRC. The network is fed with a fixed-size (224 * 224) input which will convolute and obtain the pooled activations repeatedly, then forward the outputs to the completely connected layers. The network was trained on ImageNet and incorporated various techniques of regularization, such as preprocessing of data, dropout. AlexNet triggers interest in developing different architectures of deep convolutionary neural networks.

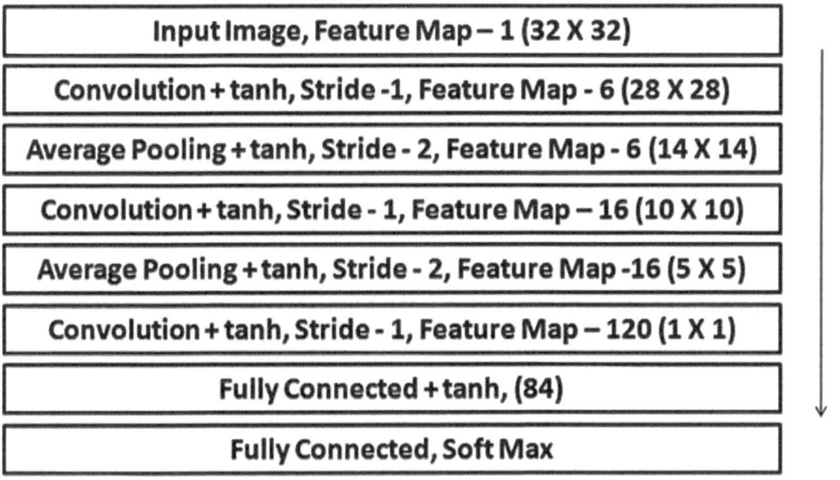

FIGURE 9.3 LeNet architecture.[1]

It is the first user of the ReLU activation function for output layer after convolutional layers and softmax. It uses max pooling instead of average pooling, dropout regularization between the FC layers, and data Augmentation for images.

In AlexNet, 11 × 11 window size forms the first convolutional layer. In the second layer, it has been reduced to 5 × 5 followed by 3 × 3 in fifth layer. Additionally, the network adds maximum pooling layers with a window shape of 3 × 3 and a stride of 2. AlexNet has convolutional channel ten times more than LeNet. The last convolutional layers are two huge fully connected layers with 4096. The early GPU has limited memory, so the AlexNet used two GPUs for storing and computing only its half of the model.

AlexNet used ReLU (simpler) activation function instead of sigmoidal activation function. The gradient of ReLU activation function in the positive interval is always 1.

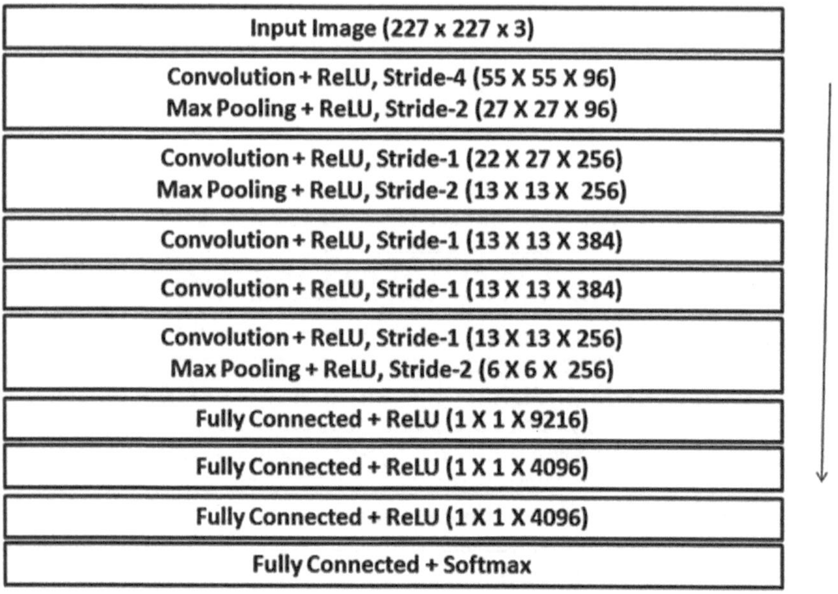

FIGURE 9.4 AlexNet architecture.[2]

iii. VGGNet

Very deep convolutional networks for large-scale image recognition owned by Visual Geometry group of Oxford University have been proposed by Simonyan and Zisserman. It was trained with 14 million images categorized to 1000 classes and generated error rate of 7.3%. The proposed net became an illustration for future deep layered structures with small filter size. AlexNet was modified by replacing multiple 3 × 3 sized filters instead of single 7 × 7 filter. A three convolutional 3 × 3 filter uses 3 × 3 × 3 = 27 parameters instead of 7 × 7 × 1= 49 parameters which also captures sophisticated features of smaller regions clearly. Deeper structure increases the nonlinearity with ReLU functions and regularizations due to repeated convolutions. The input RGB image of fixed size 224 × 224 is passed through stack of convolutional layers and filters of 3 × 3 size with one convolution stride. Spatial pooling has been done by five max-pooling layers. Maximum value-based pooling is chosen over a 2 × 2 pixel window with 2 strides. A single layer contains almost 75% parameters and the remaining in FC layers which leads to the evolution of GoogLeNet further.

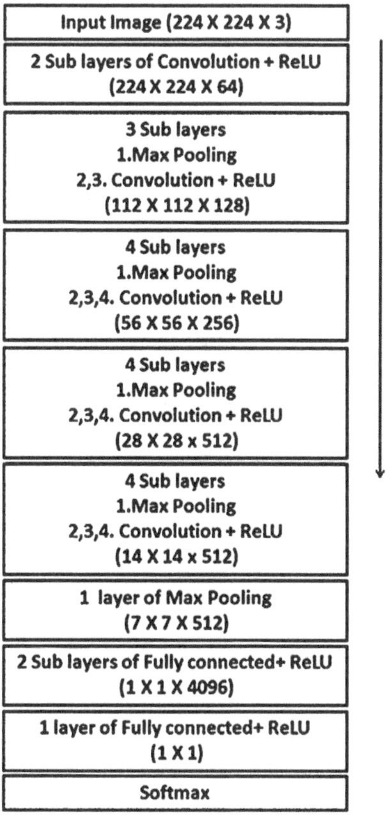

FIGURE 9.5 Architecture of VGG16.[3]

iv. GoogLeNet

An architecture modeled as network within a network referred as inception structure framed into GoogLeNet classifier. An error rate of 6.67% has been reached by Google Web with a flexible feature in selection of kernel size. Different sizes of kernels are allowed to choose along different paths such that the filtered spaces are featured sequentially with various granularity levels. Hidden features of small spaces will be magnified at larger size filters.

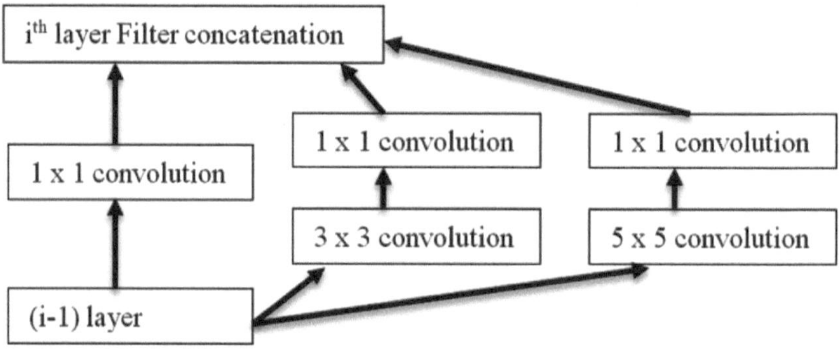

FIGURE 9.6 Basic module of GoogLeNet.

The convolution starts with 1 × 1 kernels generally as it has more depth than others to achieve good computational efficiency. The dimension reduction also helps to achieve better learning of parameters during back propagation. The network uses average pooling to cover the whole space in FC layers and hence feature extracted will be equal to number of filters.[4] The GoogLeNet outperforms VGGnet with less parameters at the cost of information loss due to average pooling. The human expert scored 5.1% as its top-5 error rate for individual model and 3.6% for the complete model. The network used Le-Net influenced CNN but introduced a foundation module. To significantly minimize the parameter size, this module is based on several 1 × 1 convolutions. Their architecture consisted of a 22-layered CNN but reduced the parameter sized 60 million to 4 million (AlexNet). The flexible feature of multi-granular decomposition with inception module (V4) led to ResNet in future.

v. ResNet 50

Originated from inception structure, ResNet is a profound network of residuals. ResNet with 152 layers generates 3.6% error and won ILSVRC competition in 2015.[5] Such a deep structure has learning limitation as convergence has not been optimized in reasonable time. The deep web structures often got vanished without adjustments, that is, the gradient becomes smaller and smaller as the model is replicated. Smaller gradients lead to cumbersome learning.

Skip link is the key innovation of ResNet. It enables the network to know the identity function, allowing the block to transfer the input without the passing through one or two weight layers. The skip link in the ResNet is called "identity." The skip link copies the input data of ith layer and supplied along with the output of i+2 layer as shown in Figure 9.7. It is done to allow the gradient flow with or without strides of padding filter. A stride of 1 does not change the size and depth of space between layers while the stride of 2 reduces the size and increases the depth of space by 2.

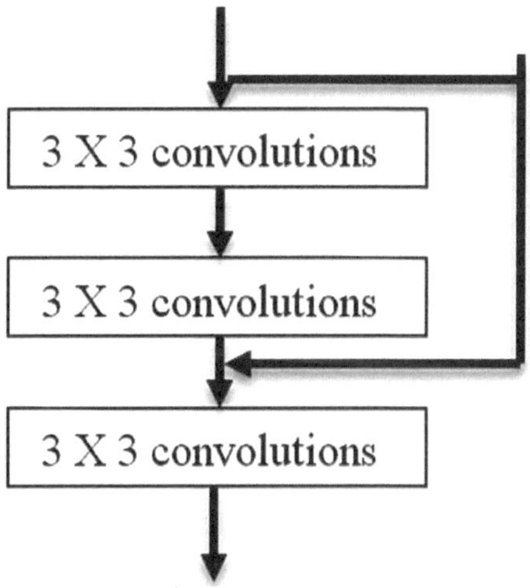

FIGURE 9.7 Residual module of basic ResNet with skip link.

Each block of ResNet is either two layers deep, or three layers deep, in small networks such as 18, 34 ResNet, and 50 ResNet. Each block of 2 layers is replaced in the net of 34 layers with this bottleneck of 3 layers which results in a ResNet of 50 layers. The short length of skip link enables easy learning during BP. The long skip link transfers the relevant features of complex space. This residual approach of learning eases seizure recognition of different granular levels (Charu, 2018). The model includes floating point operations (FLOPs) of 3.8 billion. A comparison on configuration and performance of Alexnet, Googlenet, and Resnet given in Table 9.2 shows the improvement in accuracy with added features.

TABLE 9.2 Summary of AlexNet, GoogLeNet, and ResNet Architecture.

Networks	Salient features	Top 5 performance		FLOP
		Accuracy	Error	
AlexNet 2012	Deeper network with the use of the ReLU activation function, max pooling, dropout regularization, convolutional layer fed directly to another convolutional layer, data augmentation, and 62 million parameters.	84.70%	15.3%	1.5B
GoogLeNet 2014	Wider/parallel networks with 6.4 million parameters, use of inception modules, more 1 × 1 convolutions, multiple feedback of errors and average pooling.	93.30%	6.67%	2B
ResNet50 2015	Shortcut connection with 60.3 million parameters, multiple residual modules	95.51%	3.6%	3.8B

9.6 CNN FOR SEIZURE CLASSIFICATION

Brain tumor is one of the ruinous diseases, which leads to reduce the life span or even may cause death. Sometimes, wrong observations in medical diagnosis limit the percentage of survival. Appropriate observation and proper treatment may increase the rate of survival. In our human body, brain is the most complex organ with full of tissues, neurons which activates the body and control action of body. Brain tumor is a collection of cells which is abnormal and it lost their capability. To diagnose this disease, we need an accurate and proper treatment. Before the treatment, there should be analysis of the stage of tumor. Basically, noncancerous (benign) and cancerous (malignant) tumor are the two types of tumors. Benign grows very slowly and it can be removed because it doesn't spread anywhere. But, malignant is a cancerous tumor spreads everywhere and has a rapid growth. Most common type of tumors are glioma and meningioma. For the classification and grading, computer-aided techniques are preferred. Most commonly used diagnostic technique is MRI which has more contrast on cancer tissues when compared to computed tomography (CT) images.[19,20]

Shao used a remarkable class of DL in the transfer learning rule of the studies on image classification, object recognition, and visual categorization problems. Pretrained models are developed for other application which is related to our application. TL helps to use the pretrained model of CNN for such applications. Grading of glioma form MRI brain images with the help of AlexNet and GoogLeNet by Yang et al.[22] By observing the performance of these networks, GoogLeNet manifest the AlexNet for the brain tumor grading system. Deep transfer learning place is an astonishing abnormality classification (Talo et al., 2018). Thus, brain tumor classification is the major task in computer-aided-design (CAD). The deep CNN plays a dynamic role in the medical application in this decade.

In the proposed work, three classifiers of CNN namely AlexNet, GoogLeNet, and ResNet50 are used for the brain tumor classification. The classifier categorizes the brain tumor as glioma and meningioma or nontumor. The proposed algorithm is evaluated on the Kaggle dataset which is used in many research works. From the datasets, the images are randomly divided into 70 and 30% or 80 and 20% to form the training set and validation set. The pretrained networks are used to extract the features from the MRI brain images. One of the major advantages of CNN-based classifier is there is no need to segment the tumor manually. The CNN model had five learnable layers and 3*3 size of filter size in all the layers. It provides an accuracy of about 81% for classification. To overcome these numbers of models had been applied. This proposed work provides a detailed and automated method for classification of two types of brain tumor (glioma and meningioma). This work uses a significant TL CNN model to extract the features from brain MRI images. ResNet50 is a defined as a deep network with 50 layers. The module uses kernel of sizes 1×1, 3×3, and 7×7 in its design. The FC layer consists of 1000 nodes. The ResNet50 model is the good model on the ImageNet validation set. It performed with a top-1 error rate of 20.47% and top-5 error rate of 5.25%.

ResNet50 can be used for image classification, object detection, and object visualization and also it reduces the computational expenses. The last three layers of ResNet50 are named as fc1000, fc1000-softmax, and classificationLayer-fc1000. The last three classification layers are replaced with new layers of classifier. The fine tuning of the modified ResNet50 is performed with the datasets. The learning factors for weight and bias are set as 10 at the FC layer. The main objective is to make the network learn attributes of various levels unique to the destination. The low-level attributes of the pretrained models from the original ResNet50 is supposed to be learnt. The transfer of deeply learned and sophisticated CNN could be utilized for experiments then with the MRI brain dataset. With the help of TL, the network is trained with the limited number of images. These networks help to provide the accuracy of the classifier and other parameters such as sensitivity, specificity, precision, recall, and F-score.

9.6.1 PROPOSED CLASSIFICATION FRAMEWORK

The modified and deeply tuned ResNet50 was used by the proposed work to acquire the features of MRI brain images which has tumor and nontumor. Selective representation for the trained images was good which is given the features extracted from the layer. Softmax layer in the transfer learned model classifies the data into tumor and nontumor. For the test results, the model forecast and outputs class labels.

9.6.2 PREPROCESSING INPUTS

The data from the Kaggle dataset are freely available and are frequently used for categorizing the images. It is a pool of 250 MRI images diagnosed for brain tumors. It contains 59 images corresponding to glioma, 66 images corresponding to meningioma, and remaining 132 images are nontumor. ResNet50 was primarily designed for RGB color images with an input layer of size 224 × 224 × 3. The brain MRI images in the datasets were normalized and then resized to 224 × 224 × 3. The entire dataset of 250 images was divided into 10 disjoint subsets followed by a k-fold cross-validation. Among the 10 subsets, one of them has been selected as the test input and rest of the subsets for training. Every subset has been used as test input at least once by repeating the process.

9.6.3 ALGORITHM IMPLEMENTED IN MATLAB

Step 1: Load the dataset using an image datastore to manage the data

Step 2: Use count each labels to summarize images per category.

Step 3: Visualize the images to distinguish the classification for the images.

Step 4: K-fold variation is used for the analysis. The value for k is calculated by experiments to generate a model estimate with a low range to 10, which usually results in a moderate variance. “num_folds = 10;”

Step 5: Load the pretrained model as “net = resnet50;”

Step 6: For the new learnable layer, set the weight learn factor = 10 and the Bias learn rate = 10; to get the new layers.

Step 7: Replace the last layer of the network or the output layer with the new layers.

Step 8: Train the network for the classification using training options.

Step 9: Performance of the network can be measured in terms of precision and recall, accuracy using confusion matrix (CM).

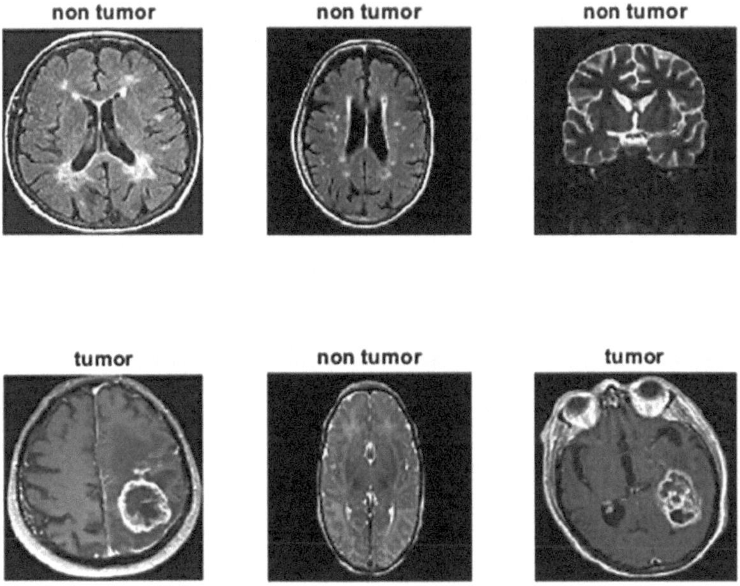

FIGURE 9.8 MRI images.

9.6.4 *CLASSIFIER SETTINGS*

The modified ResNet50 has been used for training the preprocessing set. The network’s hyper parameters have been heuristically optimized so that the loss function will get converged during training. Stochastic gradient descent momentum (SGDM) is chosen as the optimizer for its good learning rate characteristics. The learning rate is chosen as 1e-4initially for SGDM. The mini batch size is set as 8 and the number of epochs as 30. The model has been simulated 10 stretches and each run trailed a 10 times the cross-validation process. The prediction has been summarized by using CM for the chosen classifiers.

Figure 9.9 proved that the trained data reaches accuracy gradually with the chosen classifier and loss decreases with the increase in number of iterations.

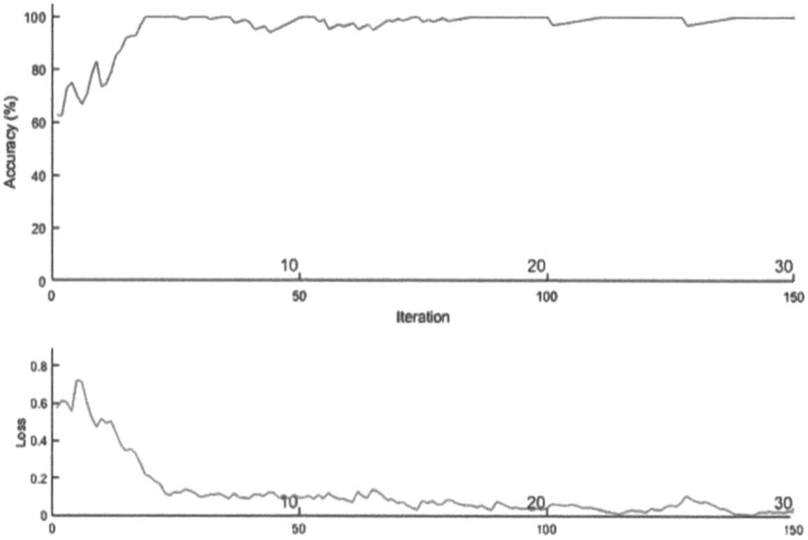

FIGURE 9.9 Accuracy and loss after training.

9.6.5 BRAIN TUMOR SEGMENTATION

Brain tumor segmentation is a segregation process of tumor pixels from normal ones. This process is still being researched using various design or architectures or classifiers or optimization techniques as it helps medical personnel by providing information on tumor growth, responses of tumor during surgery, or radiation therapy, etc. The segmented images helps to assess the boundaries and exact location of tumor in order to evaluate

TABLE 9.3 Segmented Images of CNN for Input Dataset.

Type	MRI image	Thresholded image	Segmented image
Meningioma	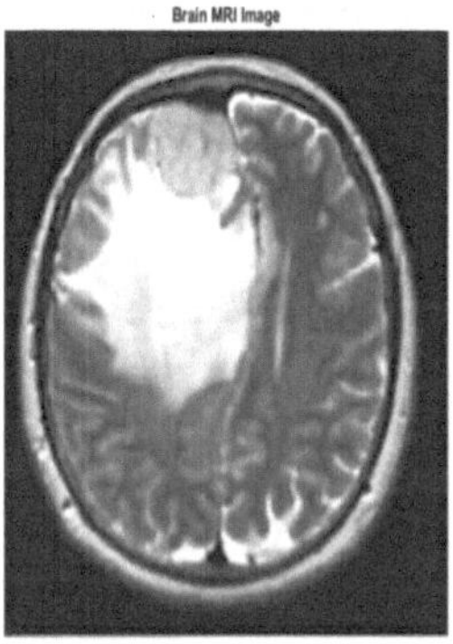	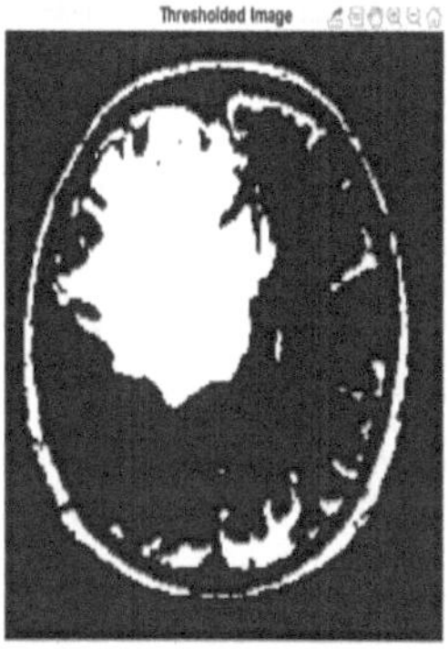	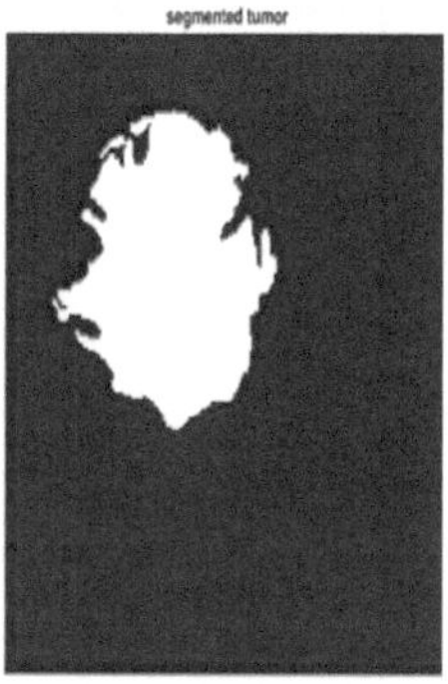
Glioma	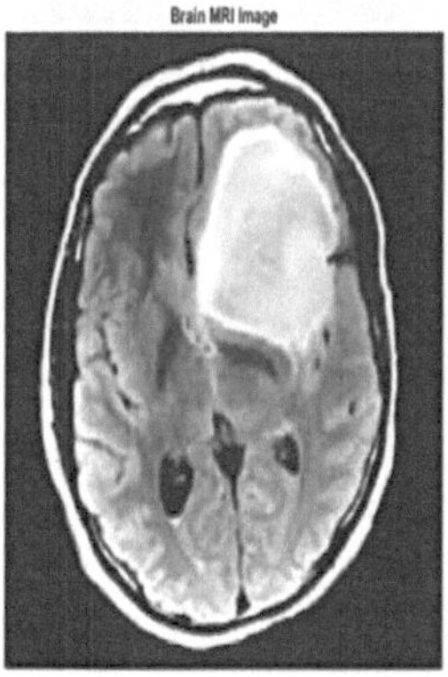	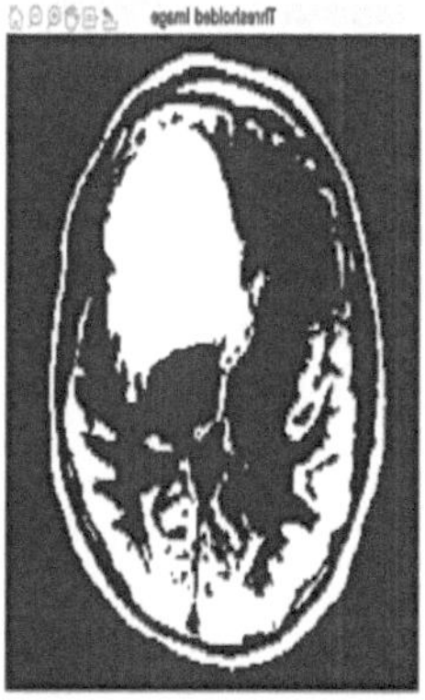	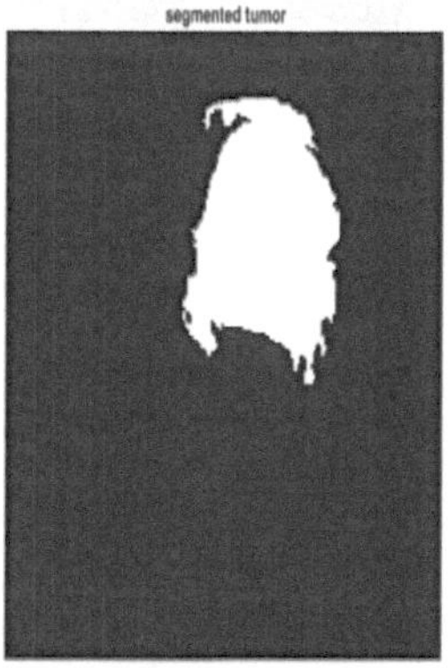
Nontumor	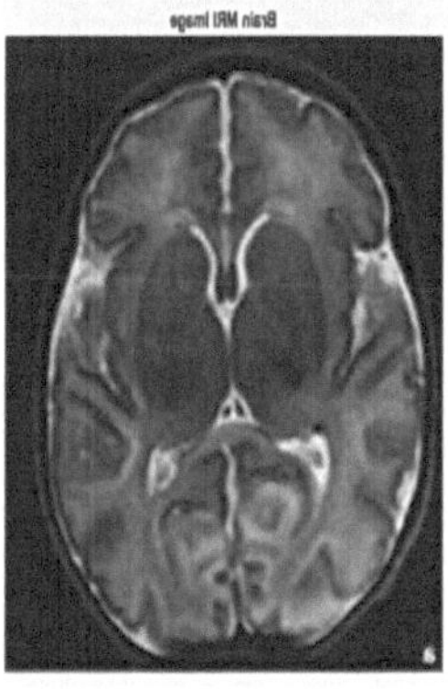	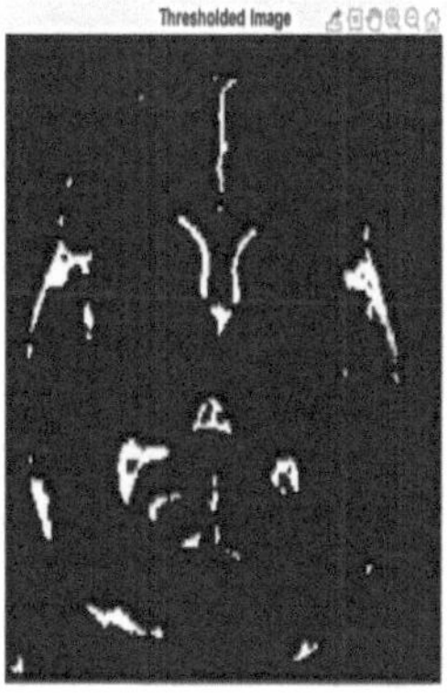	

the hostile effects of cancer cells.[24] The MRI images of the dataset are preprocessed properly and fed to the CNN layers. The segmentation of the three categories (glioma, meningioma, and nontumor) obtained at the output layer of CNN are given to the classifier layer. The output samples of CNN are tabulated in Table.9.3. The glioma sample shows that the image has undergone data augmentation process. The nontumor sample has no white pixels as it does not have tumor tissues. Metrics such as precision, sensitivity, specificity, recall, and F-score are measured using the values of confusion matrix. The correct and incorrect predictions are counted and categorized as positive (true/false) and negative (true/false) data. With the CM parameters, the performance metrics proved better performance of ResNet50 which is given in Table.9.4

TABLE 9.4 Metrics Using CM of ResNet.

Measure	Derivations	AlexNet	GoogLeNet	ResNet
Sensitivity	$P_T / (P_T + N_F)$	0.9370	0.9520	0.9683
Specificity	$N_T / (P_F + N_T)$	0.9187	0.9739	0.9685
Precision	$P_T / (P_T + P_F)$	0.9225	0.9754	0.9919
Negative predictive value	$N_T / (N_T + N_F)$	0.9339	0.9492	0.9683
False positive rate	$P_F / (P_F + N_T)$	0.0813	0.0261	0.0317
False discovery rate	$P_F / (P_F + P_T)$	0.0775	0.0246	0.0315
False negative rate	$N_F / (N_F + P_T)$	0.0630	0.0480	0.0081
Accuracy	$(P_T + N_T) / (P + N)$	0.9280	0.9625	0.9800
F1 Score	$2 P_T / (2 P_T + P_F + N_F)$	0.9297	0.9636	0.9801
Matthews correlation coefficient	$P_T * N_T - P_F * N_F$ / sqrt $((P_T + P_F)* (P_T + N_F)* (N_T + P_F)* (N_T + N_F))$	0.8560	0.9252	0.9603

Where,

P_T—True positive predicted as number of tumor pictures in a given datasets and it is indicated as "Yes,"

N_T—True negative predicted as number of nontumor picture in a given datasets and it is indicated as "NO."

P_F—False positive predicted as true, though it doesn't contain any tumor data and categorized as a Type-I error.

N_F—False negative predicted as no; however, it contains some tumor data and is categorized as Type-II error.

Table 9.5 shows the comparison of ResNet50 performance with former classifiers in terms of CM, accuracy, and time consumption for classification. Identification of true/false values has been superiorly done by ResNet compared to AlexNet and GoogLeNet. The ResNet framework with 50 layers has high sensitivity and precision than the others in the classification performance at the cost of time.

9.7 SUMMARY

The proposed work can be implemented as a fully automatic system for brain seizure detection. The image dataset is well preprocessed and the CNN-based approach utilizes the concept of TL for the feature extraction. The CNN output layer fed the segmented images to the chosen classifiers priorly used for image classification. The inference of the classification performance of all the classifiers is summarized as follows.

- The confusion matrix shows that the true positive and negative results are appropriately identified using ResNet than others.
- The ResNet framework is also sensitive and precise than the others in the classification performance.
- The performance is mainly based on iterations to train the image data and time consumed. The classifiers are iterated for 10, 50, and 60 times and analyzed for optimum accuracy. AlexNet and GoogLeNet accuracy remains the same beyond 60 iterations. But ResNet50 attained optimum accuracy at the 150th iteration. Though ResNet has comparable efficiency at 50th iteration, increasing the iteration to 150 gives better accuracy when compared to other classifiers.
- The performance analysis was taken by measuring the accuracy value of the classifiers. It is noted that the accuracy remains same for the AlexNet and GoogLeNet beyond 60th iteration. At the 150^{th}iteration, the accuracy is 92.6% for AlexNet, 96.4% for GoogLeNet, and 98.02% for ResNet50.
- As the iteration increases, the time taken to perform classification also increases. Hence, the best accuracy is attained at the cost of time as the depth of network influences the both.

It has been concluded that ResNet50 gives enhanced accuracy than other two classifiers. It is inferred that the tumor classification is performed

TABLE 9.5 Comparison of AlexNet, GoogLeNet, and ResNet for Classification.

Metrics	AlexNet			GoogLeNet			ResNet 50		
Confusion matrix	Confusion matrix: Yes: 119 48.3% | 10 1.7% | 96.6% 3.4% No: 8 5.68% | 113 44.32% | 88.64% 11.36% 95.20% 4.80% | 90.4% 09.6% | 92.60% 7.4%			Confusion matrix: Yes: 119 47.6% | 3 2.4% | 95.2% 4.8% No: 6 1.2% | 122 48.8% | 97.6% 2.4% 95.20% 4.80% | 97.6% 02.4% | 96.40% 3.6%			Confusion matrix: Yes: 123 49.2% | 4 0.16% | 98.08% 1.92% No: 1 0.04% | 122 48.8% | 97.52% 2.48% 98.4% 1.6% | 97.6% 2.4% | 98.0% 2.0%		
Specificity	91.87%			93.395%			**96.85%**		
Sensitivity	93.7			95.20			**96.83**		
Precision	0.9225			0.9754			**0.9919**		
Simulation results	Iteration	Time (s)	Accuracy	Iteration	Time (s)	Accuracy	Iteration	Time (s)	Accuracy
	10	41	60.80%	10	25	50.32%	10	62	68.70%
	50	67	70.96%	50	66	89.15%	50	98	90.00%
	60	80	92.59%	60	72	96.32%	150	158	98.02%

using ResNet50 with more iterations. While this work explored three deep architectures and TL approaches for MRI images, CT images are yet to be explored for brain tumor. The architectures for the classification of brain tumor with less iteration of other robust DNNs can be investigated in future.

KEYWORDS

- **artificial intelligence**
- **deep learning**
- **machine learning**
- **artificial neural network**
- **root mean square**

REFERENCES

1. Lecun, Y.; Bottou, L.; Bengio, Y.; Haffner, P. Gradient-Based Learning Applied to Document Recognition. *Proc. IEEE* **1998,** *86*(11), 2278–2324. DOI: 10.1109/5.726791.
2. Krizhevsky, A.; Sutskever, I.; Hinton, G. E. In *ImageNet Classification with Deep Convolutional Neural Networks*, 25th International Conference on Neural Information Processing Systems, Vol. 1 (NIPS'12); NY, USA, 2012; pp 1097–1105.
3. Simonyan, K.; Zisserman, A. Very Deep Convolutional Networks for Large-Scale Image Recognition. *CoRR* **2014,** abs/1409.1556.
4. Szegedy, C; et al. In *Going Deeper with Convolutions*, IEEE Conference on Computer Vision and Pattern Recognition (CVPR), Boston, MA, 2015; pp 1–9, DOI: 10.1109/CVPR.2015.7298594.
5. He, K.; Zhang, X.; Ren, S.; Sun, J. In *Deep Residual Learning for Image Recognition*, IEEE Conference on Computer Vision and Pattern Recognition (CVPR), Las Vegas, NV, 2016; pp 770–778. DOI: 10.1109/CVPR.2016.90.
6. Patterson, J.; Gibson, A. *Deep Learning – A Practitioner's Approach*; O'Reilly Media Publishers Inc.: USA, 2017.
7. Simonyan, K.; Zisserman, A. Very Deep Convolutional Networks for Large-Scale Image Recognition, Published as a Conference Paper at ICLR 2015.
8. Mohsen, H.; El-Dahshan, E. S. A.; El-Horbaty, E. S. M.; Salem, A. B. M. Classification Using Deep Learning Neural Networks for Brain Tumors. *Future Comput. Inform. J.* **2018,** *3*, 68–71. https://doi.org/10.1016/j.fcij.2017.12.001
9. Zhang, J.; Shen, X.; Zhuo, T.; Zhou, H. Brain Tumor Segmentation Based on Refined Fully Convolutional Neural Networks with a Hierarchical Dice Loss, 2017.

10. Alqudah, A. M.; Alquraan, H.; Qasmieh, I. A.; Alqudah, A.; Al-Sharu, W. Brain Tumor Classification Using Deep Learning Technique - A Comparison Between Cropped, Uncropped, and Segmented Lesion Images with Different Sizes. *Int. J. Adv. Trends Comput. Sci. Eng.* **2019,** *8*(6), 3684–3691. https://doi.org/10.30534/ijatcse/2019/155862019.
11. Özyurt, F.; Sert, E.; Avci, E.; Dogantekin, E. Brain Tumor Detection Based on Convolutional Neural Network with Neutrosophic Expert Maximum Fuzzy Sure Entropy. *Measurement* **2019,** *147*, 106830(1–8). https://doi.org/10.1016/j.measurement.2019.07.058
12. Seetha, J.; Raja, S. Brain Tumor Classification Using Convolutional Neural Networks. *Biomed. Pharmacol. J.* **2018,** *11*(3), 1457–1461.
13. Nadeem, M. W.; Al Ghamdi, M. A.; Hussain, M.; Khan, M. A.; Khan, K. M.; Almotiri, S. H.; Butt, S. A. Brain Tumor Analysis Empowered with Deep Learning: A Review, Taxonomy, and Future Challenges. *Brain Sci.* **2020,** *10*, 118.
14. Ari, A.; Hanbay, D. Deep Learning Based Brain Tumor Classification and Detection System. *Turk. J. Electr. Eng. Comput. Sci.* **2018,** *26*, 2275–2286.
15. Ucuzal, H.; Yaşar, Ş.; Çolak, C. In *Classification of Brain Tumor Types by Deep Learning with Convolutional Neural Network on Magnetic Resonance Images Using a Developed Web-Based Interface*, 3rd International Symposium on Multidisciplinary Studies and Innovative Technologies (ISMSIT), Ankara, Turkey, 2019; pp 1–5. DOI: 10.1109/ISMSIT.2019.8932761.
16. Amin, J.; Sharif, M.; Gul, N.; Yasmin, M.; Shad, S. A. Brain Tumor Classification Based on DWT Fusion of MRI Sequences Using Convolutional Neural Network. *Pattern Recognit. Lett.* **2020,** *129*, 115–122.
17. Das, Sunanda; Aranya, O. R. R.; Labiba, N. N. In *Brain Tumor Classification Using Convolutional Neural Network*, 1st International Conference on Advances in Science, Engineering and Robotics Technology, 2019 (ICASERT 2019).
18. Abiwinanda, N.; Hanif, M.; Hesaputra, S. T.; Handayani, A.; Mengko, T. R. Brain Tumor Classification Using Convolutional Neural Network. In *World Congress on Medical Physics and Biomedical Engineering 2018, IFMBE Proceedings 68/1,* 2018; Lhotska, L., et al., Eds.; https://doi.org/10.1007/978-981-10-9035-6_33.
19. Yousefi, M.; Krzyiak, A.; Suen, C. Y. Mass Detection in Digital Breast Tomosynthesis Data Using Convolutional Neural Networks and Multiple Instance Learning. *Comput. Biol. Med.* **2018,** *96*, 283–293.
20. Gu, Y.; Lu, X.; Yang, L.; Zhang, B.; Yu, D.; Zhao, Y.; Thou, T. Automatic Lung Nodule Detection Using a 3D Deep Convolutional Neural Network Combined with a Multi-Scale Prediction Strategy in Chest CI's. *Comput. Biol. Med.* **2018,** *103*, 220–231.
21. Shao, L.; Zhu, F.; Li, X. Transfer Learning for Visual Categorization: A Survey. *IEEE Trans. Neural Netw. Learn. Syst.* **2015,** *26*(5), 1019–1034.
22. Yang, Y.; Yan, L. F.; Zhang, X.; Han, Y.; Nan, H. Y.; Hu, Y. C.; Ge, X. W. Glioma Grading on Conventional MR Images: A Deep Learning Study with Transfer Learning. *Front. Neurosci.* **2018,** *12*, 804. https://dx.doi.org/10.3389%2Ffnins.2018.00804
23. Talo, M.; Baloglu, U. B.; Acharya, U. R. Application of Deep Transfer Learning for Automated Brain Abnormality Classification Using MR Images. *Cognitive Syst. Res.* **2019,** *54*, 176–188. https://doi.org/10.1016/j.cogsys.2018.12.007

24. Rajasekaran, K. A.; Chellamuthu, C. Advanced Brain Tumour Segmentation from MRI Images; *In High-Resolution Neuroimaging*; Intechopen Publications, 2018. https://doi.org/10.5772/intechopen.71416.
25. https://www.simplilearn.com/deep-learning-algorithms-article.
26. https://cs231n.github.io/convolutional-networks/
27. https://d2l.ai/chapter_convolutional-neural-networks/conv-layer.html
28. https://towardsdatascience.com/a-comprehensive-guide-to-convolutional-neural-networks-the-eli5-way-3bd2b1164a53
29. https://www.pyimagesearch.com/2016/08/01/lenet-convolutional-neural-network-in-python/
30. https://engmrk.com/lenet-5-a-classic-cnn-architecture/

CHAPTER 10

A Proactive Improvement Toward Digital Forensic Investigation Based on Deep Learning

VIDUSHI*, AKASH RAJAK, AJAY KUMAR SHRIVASTAVA, and ARUN KUMAR TRIPATHI

KIET Group of Institutions, 201206 Ghaziabad, India

Corresponding author. E-mail: vidushi.mtech@gmail.com

ABSTRACT

This chapter presents a proactive improvement toward digital forensic investigation based on Deep Learning. Digital Forensics has witnessed exponential growth in the last few years. It builds the trust of users toward digital practices. However, efficient forensics technology is required for adequate security. Recently, deep learning has been widely used for accomplishing digital forensic tasks more accurately. It provides a much more efficient means of digital forensics. In this chapter, the various learning methods, along with their respective merits and demerits, are discussed. To accelerate the investigation process using digital forensics, the research deeply analyses the convolutional network with its available applications, including probe study of images and videos. The authentication process using image recognition and the exigency of the deep learning model is discussed. Moreover, the smart Internet of Things (IoT) concept, architecture, and the applicability of network forensics in IoT devices are introduced in the chapter.

Deep Learning in Visual Computing and Signal Processing. Krishna Kant Singh, Vibhav Kumar Sachan, Akansha Singh & Sanjeevikumar Padmanaban (Eds.)

10.1 INTRODUCTION

Throughout the years, in the fast pace, dramatic growth has been exhibited in digital forensic[1,2,3] and remarkable success in deep learning techniques.[4,5] The consciousness and trust toward the scientific knowledge of society enormously increased that also attracts the scientists to use scientific methods to investigate legal mess. The science used to tackle such legal issues is termed as forensic science.[6,7] Forgeries[8,9,10] have extremely increased that can rampant and disturb stability in person's life. The thorough investigation through the digital forensic appears as the suitable, trusty solution. Therefore, researchers are paying their full attention to identify the forgeries by investigation using digital forensic. The forensic identification can be performed by matching the doubtful activities with real one. Recent trend in forensic science is abundantly applied through the deep analyses of the questioned evidences. The forgery question can be raised on image, video, and many more. There are several state-of-art techniques present such like JPEG compression,[11] and others. However, currently image recognition[12,13] is widely used and plays a vital role in digital forensic area. The proper assessment, learning, and evaluation of the evidences from the image can be successfully accomplishing using image recognition. This learning would be further helpful in corroboration and authentication.

Image recognition is helpful and has paramount important applications in forensic investigation[14] such as handwriting or character scanning,[15] mechanism for face recognition,[16] estimation of age with the aid of radiological data, analyses of steganographic images,[17] judgment of hyperspectral pictures,[18] detailed examination of acquisition devices, fingerprint investigation,[19,20] and the study of images captured using the camera. Along with the images, video anatomization that includes CCTV footage is the modern age and well-known applications of forensic science. Recognizing images, videos need deep sincere study, it is a hard challenge but a fruitful step toward forensic. Researchers are giving their efforts to recognize or learn minute to minute points for forgery detection.

Over the recent past years, continuous changes are coming in the society with the movement in the internet era. Now, the direct physical face-to-face interaction lows down in society and turns to the online interaction such like video conferencing, chat, image sharing, data sharing, and many more. In the direct physical interaction or in pre-internet era, the requirement of digital authentication[21,22] investment technique is very little, almost nil. As with the enhancement in the technology, physical interaction lows down, drastically raise the demand and necessity of authenticity in market. The momentous authenticity requirement wakes up the researchers toward the serious security concern. The security issues faced by the society can lead to great damage. The researchers are working in this area to overcome these security challenges. Worldwide scientists are giving their efforts to improve and ensure the reliability and to overcome the security threats.

The main challenge comes in front of researchers is handling and analyzing the enormous volume of data with accuracy. Analyzing illegal movements is a tedious task as to recognize minute key points from the available huge data. In order to find out small section within the massive information needs intelligent work because small portion can be only relevant to unlawful event. The forensics ultimate motive is to find out the suspicious activity and the one who is behind that suspicious event. To turn the suspicious one into the real fact as per the court law with reasonable points of either faulty or innocence is the target. Experts of forensic attempt to analyze the suspicious situation, which needs lots of effort and based on analyses, and take the appropriate action. In recent decades, with the drastic increase in the demand of digital forensic also enhances the pressure on the forensic expert. This demand puts forward the censorious concerns related to forensic techniques validity and how much these are reliable that can be further proved in court.

The presence of enormous bulky data should be investigated to control the forgery; a shift to digital forensic becomes inevitable. This requires some automatic intelligent learning mechanism for high level of computational analyses, vision, and recognition. These computational learning approaches provide the machines with the required recognition and analyses ability. Along with it, in respect to provide the procedures related to scientific learning methods can support the huge bulky data volume. The capability of the computational learning approaches must be able to have expert knowledge that helps to avoid the human intervention and overcome the human burden.

Recently, machine learning,[23,24,25] consciousness is continuously spreading with fast pace in the public domain and parallelly also exploded in research. Researchers are applying various learning algorithms in diverse areas, especially in prediction or classification task. These learning approaches show their fruitful outcomes even while dealing with complex and hard challenges. The analytical capability of learning models is suffusing the world because of getting the paramount success. In order to think about machine independence from human and able to work without human intervention seems obscure before the implementation of learning models. Still the research work is going on because of the existence of some uncertainty. This uncertainty opens the research gates for further improvement.

Machine learning is the branch of artificial intelligence[26,27] and deep learning is the machine learning subset. Deep learning involves the mechanism of artificial neural network. The inspiration for this network comes from the natural biological working of human brain.[28] This artificial implementation of neural network is capable of handling not only linear, structured data but also can tackle effectively nonlinear unstructured scattered data. The development toward artificially implemented neural network[29] improves the process to predict, identify, as well as recognize that further assist in taking important decisions. This learning furnishes the researchers with the clear vision and positively turns in the futuristic research. The researchers intend to use the different models of artificial network to develop an efficient and easy-to-use model with maximum accuracy.

The ultimate motive of the research is to use the learning approaches for successful accomplishment of digital forensic such as authentication through face recognition, fingerprint judgment, foot printing, and many more. There are number of different learning models present that have their own practical areas.

Convolutional neural network (CNN)[30,31] is one of the successful models that comes under deep learning. The remarkable benchmark achievement is shown by convolutional network, especially in the discipline of computer vision. The convolution allows the resolution control explicitly and at which computation of feature responses are done. To work with images, all the features including the low level are processed such as edges, boundaries, and curves. This leads to build more abstract notion through convolutional layers series. This network can be used to enhance security through the authentication like face, footprint, or fingerprint.

Similarly, recurrent neural network[32,33] is another model of deep learning that also has memory to store the previous computations. It is mainly useful in predicting the sentence next word based on previous word appeared in time step. In the same way, other neural networks are available and have their own applications. To enhance and promote the use of digital forensic, convolutional network can be a good solution. It can recognize the unauthenticated person effectively and efficiently.

The research major contribution is toward the use of deep learning approach for accomplishing the digital forensic task more accurately and in a better efficient manner. The study detailed the various learning methods along with their respective merits as well as demerits. In the way to accelerate the investigation process using digital forensic, the research deeply analyzed the convolutional network with its available applications that include probe study of images as well as videos. The authentication process using the image recognition and the exigency of deep learning model is discussed in the document. Moreover, the smart Internet of Things (IoT) concept, architecture, and the applicability of network forensics in the IoT devices are introduced in the chapter.

The remaining chapter is organized into different sections. The following Section 10.2 provides background of related work with respect to learning approaches, forgery, and other related areas. Next, the digital forensic investigation concept is discussed in Section 10.3. Moreover, the upcoming Section 10.4 explains the concept deep learning approaches. Further, Section 10.5 introduces the convolutional model in detail. Additionally, the Section 10.6 detailed the experimental work using CNN and two known image datasets, namely MNIST and Dogs vs Cats. Finally, Section 10.7 concluded the complete research work and discusses the future-plan.

10.2 BACKGROUND

Vishwakarma et al.[34] emphasis is on the reidentification of person using Gaussian Models having Multilevel. This work proposed a descriptor model of new features using Gaussian distribution model multilayer framework on pixel attributes such as color moments and space values, Schmid filter responses and gradient information. Also, the research analyzed several available metric learning models efficiency using this descriptor. Finally, the study demonstrates the performance of proposed descriptor better than others.

Lanh et al.[35] focus on the survey of forensic methods for images through digital camera. The study briefly introduces the internal significant processing stages of digital camera. Along with, several methods are reviewed for digital camera source identification as well as to detect forgery.

Nickolaos et al.[36] provides the various mechanisms of deep learning and forensics for IoT botnets. A review of diverse mechanisms of deep learning with forensics is employed in this study to investigate their applicability and botnets for IoT environments. Further, the chapter provides a new IoT definition with the taxonomy of solutions for network forensic.

In addition, the document investigates deep learning applicability for network forensics, and inherent challenges in applying techniques of network forensics to the method called IoT.

Stern et al.[37] of the research focus on the age estimation by proposing an automatic multifactorial method for age estimation based on hand, teeth, and clavicle MRI data. The research uses convolutional network and trained it on a dataset having 322 subjects. Finally, the study overcomes the multifactorial methods limitations that the forensic practice currently used.

Rehman et al.[38] use the approach of deep learning to identify writer by visual features automatically. The chapter uses text line handwriting images written in Arabic as well as English languages for writer identification by applying deep transfer CNN. The different CNN freeze layers are evaluated that affect the writer identification rate. This chapter uses transfer learning as a pioneer work by using base dataset called ImageNet and target dataset called QUWI. To decrease over-fitting chance different techniques of data augmentation are applied using target dataset text-line images. The highest accuracy 92.78% is realized in English using Conv5 freeze layer, in Arabic 92.20%, and Arabic-English combination achieved 88.11%.

Christian et al.[39] aims to recognize photo-sketch face using the architecture based on deep learning. The issues like the availability of limited sketch images are tackled by tuning the pretrained state-of-art models from recognizing face photo to recognizing photo-sketch face using transfer learning. This uses morphable models having three dimensions for new images synthesis and expanding training data artificially, in testing phases synthetic sketches are used for performance improvement. As compared with leading methods, proposed framework reduces 80.7% error rate in viewed sketches and lowers the rank for mean retrieval by 32.5% for forensic sketches of real world.

Xinghao et al.[40] focuses on the basis of domain knowledge camera identification using multitask learning and evaluating the proposed framework on three tasks such like brand, model, as well as device-level identification by using original along with manipulated images. The chapter proposed method classification output is comparatively much effective together and more robust.

Jianyu et al.[41] focuses and demonstrates forensic investigation through digital medium. The study performs the evidence analyses along with extraction on the basis of video. The motive of this research is to provide the assistance to the forensic investigation by developing advance techniques of video analyses for digital forensic. The chapter proposes a framework for forensic analyses of video to employs efficient algorithm for enhancing video/image for analyzing the low-quality footage. In order to provide the assistance for forensic analyses based on video, deep learning technique is proposed for object detection with tracking to identify important suspects from footage.

Khan et al.[42] of this research provides a review that shows the modern trends for analyzing the hyperspectral image. The review work demonstrates the analyses and various modern applications of hyperspectral image. Due to its modern and wide applications such like forensic examination, food inspection, image surgery, remote sensing, and other; it gains immense research interest. Moreover, the study also presents the hyperspectral image used to examine the forgery detection in questioned or suspected documents by using deep learning.

Hosler et al.[43] address the problem of limited standard databases of digital video; these databases are essential for development and evaluation of state-of-art algorithms for video forensic. For this need, the study presents the video-ACID database, ACID is authentication and camera identification database. This database has more than 1200 videos taken from 46 devices that represent unique 36 camera models. Table 10.1 shows the previous research briefly.

Table 10.1 depicts the analysis of fruitful research performed by scientists related to forensic analysis. Table 10.1 indicates the background past studies review work that would be helpful for future research.

10.3 DIGITAL FORENSIC INVESTIGATION ANALYSIS

Digital forensic[44,45] deals with the digital devices investigation to find out the suspicious events. All the present devices which can carry digital data are the digital devices, not only computers but the digital camera, watch, etc. comes under the digital device's category. As the involvement of digital devices are drastically increases in human life and touches the day to day activity, in parallel fashion security concern also seen. With the movement of criminal activities toward the cyberspace, criminals start exploiting the system through illegal capturing of information. These

TABLE 10.1 Previous Research Analysis.

S. No	Paper title	Authors	Technology used	Problem mentioned	Dataset used	Conclusion
1	A deep structure of person reidentification using multilevel Gaussian models.[34]	Dinesh Kumar Vishwakarma, Sakshi Upadhyay.	Multilevel Gaussian models	Person reidentification	VIPeR, QMUL GRID, PRID4 50S, CHUK 01	Proposed descriptor is more robust.
2	A survey on digital camera image forensic methods.[35]	Tran Van Lanh, Kai-Sen Chong, Sabu Emmanuel, Mohan S Kankanhalli.	Survey about forgery detected n and identification of source	Image forensic	NIL	Theoretical Examination of methods for forgery detection and identification through camera.
3	Forensics and deep learning mechanisms for botnets in Internet of Things: A survey of challenges and solutions[36]	Nickolaos Koroniotis, Nour Moustafa, Elena Sitnikova.	Deep learning	A survey for IoT botnets	NIL	A review to investigate the IoT botnets.
4	Automatic age estimation and majority age classification from multifactorial MRI Data[37]	Darko Stern, Nicola Giuliani, Martin Urschler.	Deep convolutional neural network (DCNN)	Age estimation	X-ray image dataset	Overcome multifactorial methods limitations.
5	Automatic visual features for writer identification : a deep learning approach[38]	Arshia Rehman, Saeeda Naz, Muhammad Imran Razzak, Ibrahim A Hameed.	Deep Learning CNN approach, and SVM.	Person identification using writing	ICDAR 2013, CVL database.	Got highest accuracy with layer freeze Conv5: English:-92.78%, Arabic:-92.20%, both languages:-8811%

TABLE 10.1 *(Continued)*

S. No	Paper title	Authors	Technology used	Problem mentioned	Dataset used	Conclusion
6	Forensic face photo-sketch recognition using a deep learning-based architecture[39]	Christian Galea, Reuben A. Farrugia.	Deep convolutional neural network (DCNN)	Forensic recognition of face photo-sketch.	PRIP-HDC dataset	Error rate reduced by the proposed framework: viewed sketches: 80.7%, forensic sketches for real world: 32.5%
7	Camera identification based on domain knowledge-driven deep multitask learning[40]	Xinghao Ding, Yunshu Chen, Zhen Tang, Yue Huang.	Deep learning	Camera identification for image forensic	Dresden database	A method for camera identification is introduced on the basis of multitask learning
8	Video-based evidence analysis and extraction in digital forensic investigation[41]	Jianyu Xiao, Shancang Li, Qingliang Xu.	Deep learning	Analysis of video-based evidences for forensic investigation	NIL	Framework for forensic investigation based on digital video is proposed.
9	Modern trends in hyperspectral image analysis: a review[42]	Muhammad Jaleed Khan, Hamid Saeed Khan, Adeel Yousaf, Khurram Khurshid, Asad Abbas.	Deep learning	Analysis of hyperspectral images.	Barrax dataset.	A review paper for analysis of hyperspectral images.
10	The video authentication and camera identification database: a new database for video forensics[43]	Brian C. Hosler, Xinwei Zhao, Owen Mayer, Chen Chen, James A. Shackleford, Matthew C. Stamm.	Deep learning (a new database proposed).	Video forensic database	Video-ACID (authentication and camera identification) database.	Video ACID (a new standard database is proposed).

illegal activities need some laws termed as digital forensic. In the present time, the examination of digital devices and forensic analysis gains the utmost importance. It also acquires the researcher's keen attention and opens the door for new innovative ideas to overcome the security concern.

There are several digital forensic subfields; each provides investigating specialized techniques for security incidents in IT domain. Figure 10.1 shows several digital forensic subfields.

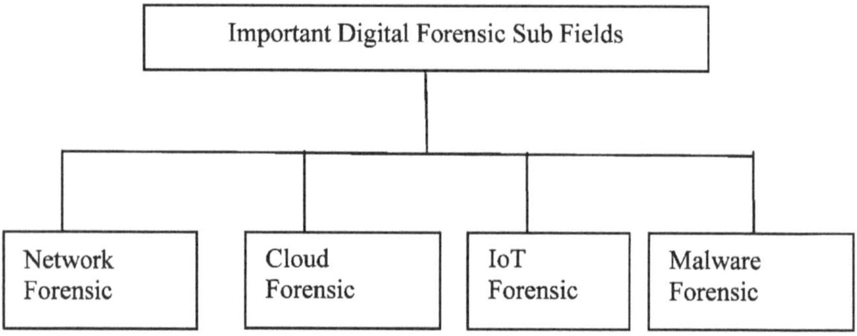

FIGURE 10.1 Several digital forensic subfields.

Figure 10.1 depicts the important subfields of digital forensic. The brief details of these fields are as follows:

- Network Forensic:[46,47]
  This forensic subfield can be emerged to identify, analyze, and assemble evidences against the malicious activities for taking the required legal action. The attackers use the internet and network technologies as a bridge to attack, such as data theft. To capture the evidences by this forensic technique is a tough task because evidences are short lived in the network. Intrusion detection system (IDS) is the well-known tool that deals with the security threat in network. In order to recognize the malicious traffic patterns in the network, training, and then validation of IDS system is done. Honeypots is the important network security famous tool.

- Cloud Forensic:[48]
  This subfield of digital forensic is dealt to investigate the concern incidents of security that may be present in the cloud. It can be termed as cross-disciplinary sector, like device and network forensic

analyses. It is responsible to oversee the committed crime if happens in the cloud. Cloud forensic can be defined as digital forensic applied on cloud computing. The network of cloud computing is huge and expended all over the world. It has numerous volumes of data where digital forensic is required to accomplish the proper criminal investigation.

- IoT forensic:[49]
  It is a new forensic field, comes in market because of the availability of IoT devices in market. The process of IoT forensic analyses the IoT devices, and layers to investigate the serious criminal issues. Along with the rapid increment of IoT devices in the society, a serious security concern also comes. The security threat arises because of the network connectivity between the devices. This security concern generates the need of forensic investigation to control the offences. These devices traces open the way to perform forensic investigation on the highly available data rises speedily from these devices.
- Malware forensic:[50]
  The malware forensic can be simply defined as the forensic discipline to perform investigation and analysis of malware such like virus, worm, and many more to determine their impact. Malware also termed as malicious program which has the negative intention to theft the user data or can attack the system.

Digital forensic can be defined as a process of analyzing digital devices to get the valid evidences regarding questioned situation. The various steps taken by this process are as follows:

- Identification: It is the first step in which identification purpose, suspected evidence, and other basics are identified.
- Preservation: It preserves the digital devices so that no one can make any change to it.
- Analysis: The investigation is performed in this phase by examining and analyzing the suspected event and digital devices.
- Documentation: In this phase, visible data record is created that further helps in recreating crime scenario to review.
- Presentation: The conclusion summary with explanations is presented in this last step.

Digital forensic is a science discipline which is used to handle questioned and suspected events. The various applications, advantages, and the related demerits are mentioned in Table 10.2.

TABLE 10.2 Digital Forensic Basic Concepts.

Objective	Application	Advantages	Disadvantages
It helps to analyze and preserve the crime-related evidences.	Fraud investigation	Provide integrity to the system.	Costlier than traditional methods.
Try to postulated crime motive and culprit behind it.	Theft related to intellectual property	Effective method to enhance digital work.	Limited technical resources and knowledge.
Quick and accurate evidence identification.	Forgeries matters	It helps to control the malicious activity.	Producing authenticated evidences is a tough task.
Prevent digital evidences to be corrupted.	Digital camera and other digital devices investigation.	Try to get the factual information in less time.	Video-audio recording needs high level of hardware software technology.
Attempt to recover deleted or tempered files.	Analysis of hyperspectral image, handwriting, etc. to get the crime and criminal-related information.	No need to be physically present to track the cybercrime worldly.	Digital information can be hacked and altered.
Generating valid report	Video authentication.	Save the precious time.	Preserving the evidences is also, a challenge.

Table 10.2 briefly explains the main characteristics of digital forensic. This investigation technique has its own merits and demerits as mentioned in Table 10.2. In the current era, the use of digital devices, IoT devices is drastically increasing in the real life, so it becomes important to study and analyze them properly. The public places like restaurants, malls, etc. have digital camera facility, if any criminal or unlawful activity happens then using forensic investigation through these available digital devices like camera, actual and factual examination can be done successfully to judge the suspected incidence.

Presently, devices are communicating with each other through the network in which devices are connected using internet. To enter the malicious code in the data packets while moving the data from one to another is not a tedious task for illicit persons. To handle such type of illegal actions many forensic areas are available such like network forensic, malware forensic, and others, some of the important ones are briefly explained above.

Digital forensic is the crucial scientific investigation which plays a vital role to accelerate the research speed. As the security threat can be a big barrier in research or even can completely stop in taking benefits of new advance technologies. For instance, authentication using image recognition can be used as a tool against unauthorized access.

10.4 DEEP LEARNING

Deep learning is the present most popular fast-growing learning approach. As it is machine learning sub-domain comes under area of artificial intelligence. Deep learning is not only capable to deal with linear data but also has the capacity to solve the complex, nonlinear, and unstructured dataset. This advanced learning approach attracts and influences the researchers and industries in the world to work with these techniques. These approaches show the significant improvements in many domains. Several important domains which are positively impacted by using the deep learning approaches are shown in Figure 10.2.

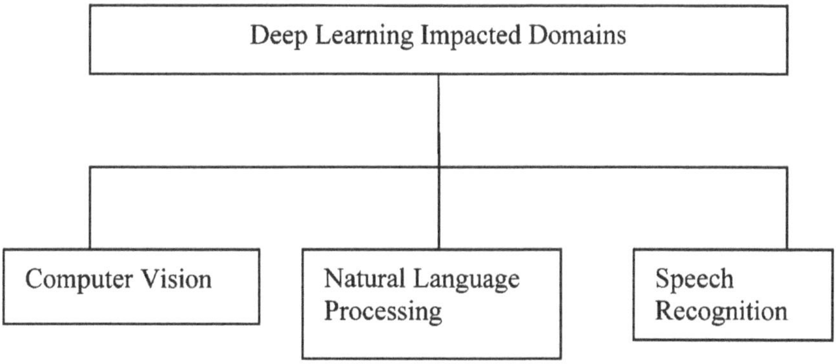

FIGURE 10.2 Several domains positively impacted by deep learning.

Figure 10.2 depicts some of the important domains whose performance is significantly improved by using deep learning. The learning has various approaches to handle different types of crucial tasks. However, the basic workflow uses artificial neural network[51,52] levels. This neural network has multiple layers consisting of neurons. These layers can look like input-hidden/s-output layer combination. To receive the input from the outside world input layer is used. For performing the desired operations on the received input from input layer numerous or at least one hidden layer/s is used, and finally to generate the predicted result output layer is used. Artificial network tries to work in the naturally intelligent human brain style.

All the artificial models of this neural network are trained by using the concept of weight updation. The basic working of this model is explained in Figure10.3.

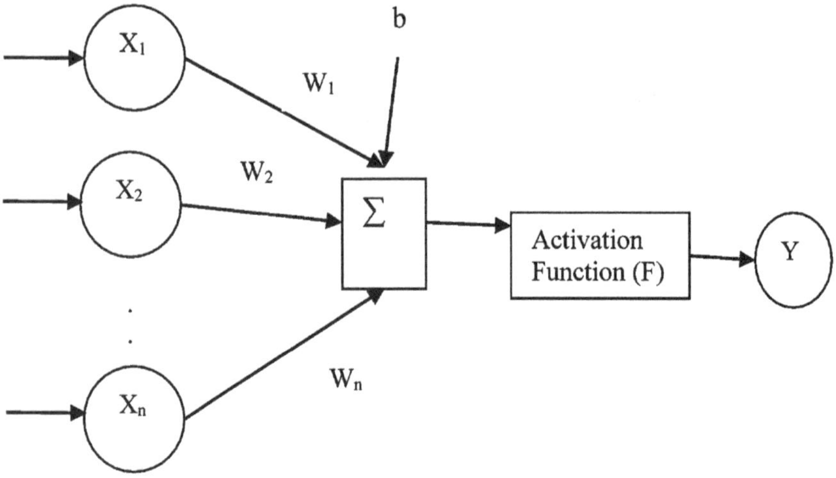

FIGURE 10.3 Basic architecture of artificial neural network.

The basic artificial model of neural network is depicted in Figure 10.3, where x1, x2, … xn are inputs, w1, w2, … wn are respective weights, b is the bias, ∑ is the summation of the products of inputs and the respective weights, F is the activation function, and finally Y is the output of the network.

The basic functionality is explained using eq 10.1.

$$Y = F(b + \sum^{n}{}_{0=1} x_i w_i) \tag{10.1}$$

Equation 10.1 describes the basic working of artificial model of neural network. In the abovementioned eq 10.1, Y shows the output of the model, F as the activation function, xi as the inputs to the model, wi are weights, n shows the neurons number.

There are various techniques present for different types of problems, but all the techniques follow the basic working of artificial model of neural network as explained above. All the models are layer based where each layer has many neurons connected with other layer neurons form a complex network.

These artificial neural models are capable of learning and on that basis can perform various tasks. These tasks need high level of computation because of the presence of nonlinearity, and unstructured behavior in data. The common crucial tasks that can be performed by them are mentioned in the Figure 10.4.

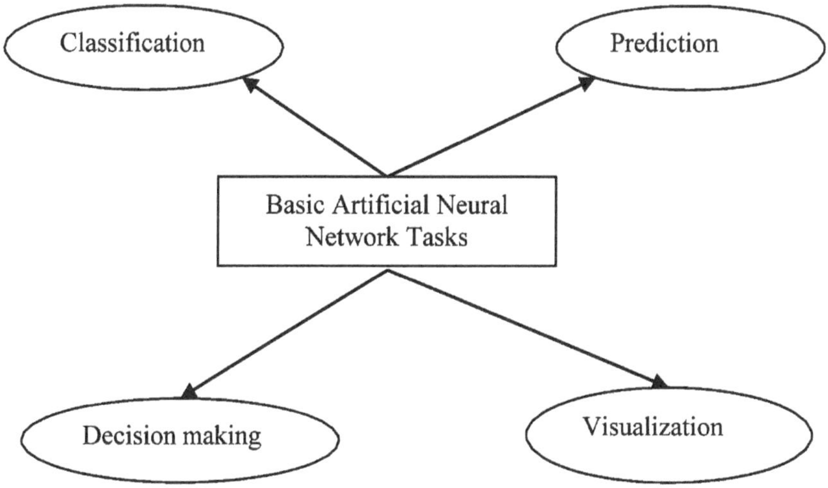

FIGURE 10.4 Artificial neural network different tasks.

Figure 10.4 mentioned all the basic tasks that have to be performed by various neural network models. To accomplish these jobs in different scenarios, different approaches are present. The flexible, powerful, and widely used approaches are as follows:

- Convolutional neural network: This is one of the most well-known artificial networks and capable to give the outstanding performance, especially with images. This network has achieved paramount success when dealing with images, text, video, and other classification or recognition jobs.

- Recurrent neural network: The specialty of this network is that its output dependency is on both present inputs as well as previous neuron state step. The performance of this network is to tackle natural language processing jobs efficiently and be able to know next word of a sentence.
- Feed forward neural network: The simplest artificially implemented neural network form where working is performed in only single way or direction. The back propagation method is not implemented in this neural network. This network receives input, performs operations, and then gives output.
- Long short-term memory: It is a particular recurrent network which is mainly modeled for temporal sequences with their respective dependencies for long range.
- Multilayer perceptron: It has a complicated network, where inputs received are processed through more than one hidden layer, and then finally output is generated. It works in both the directions called forward and backward propagation.

10.5 CONVOLUTIONAL NEURAL NETWORK

This is one of the artificially implemented neural networks whose specialty is toward handling the image analysis with recognition problem. This network gets the inspiration from the biological brain part cerebral cortex region called visual cortex. The main motive is to retrieve the minute image information and analysis of the patterns so that on that basis it is able to recognize the new image. Application of this network is widely spread in classifying and recognizing objects, faces, images, and many more.

The basic artificial network has input-hidden/s-output layers. Including all these layers, it also has several important layers. Input passes through each layer, processed accordingly, and then final output is generated. Basically, layers are used for three important functions that can be stated as "ReceiveInput-FeatueLearning-Classification." To complete the mentioned task successfully and efficiently, the complete architecture is divided into the number of layers and individual layers are composed of neurons. The basic description of each layer is as follows:

- Input layers: This layer gets the input dataset; in case of convolution network input dataset can be in the form of text, image, video, etc.
- Convolution layer: To extract the input dataset features, this is the first layer. It learns the image features for preserving the pixels relationship. The input is convolved by this layer and then further passes to the next connected layer. This is almost like neuron response in the organization of visual cortex. This layer has some hyper-parameters like filter/kernel.
- Pooling layer: It performs dimensionality reduction and reduces the parameters, if required. Because of this reduction functionality less computation is needed in data processing. The pooling can be of different types, for instance max pooling that result with maximum value using kernel, average pooling results with average value using kernel.
- Fully connected layer: The flattened matrix in the vector form is given to this fully connected layer. Fully connected means all the neurons of the layer are connected through their respective link to the neurons of the next layer. It is like a small neural network. Finally, this layer classifies the input and assigns them to their respective classes.

Goes through all the mentioned layers, the final output class is predicted. The intention is to achieve the maximum accuracy by classifying the input to the correct class. All the layers perform their task one by one and one layer transmits their output to the next connected layer. The communication between the layers and the workflow of the convolution network is shown in Figure 10.5.

Figure 10.5 mentioned above explains the convolution network working. Along with all the layers, this neural network also uses some functions such like RELU and Softmax, as mentioned in Figure 10.5. These functions are as follows:

- RELU is defined as an activation function. It expands as Rectified Linear Unit. This function can be explained using eq 10.2.

$$y = \max(0,x) \tag{10.2}$$

Equation 10.2 is linear for every "x" value more than 0. Graphically, Figure 10.6 explains it.

It is the simple activation function whose value is zero for negative and x for positive. Its math is easy to understand.

- Softmax function is an activation function. This function input is k real numbers vector and follows the method of probability distribution. Its output varies in between 0 and 1.

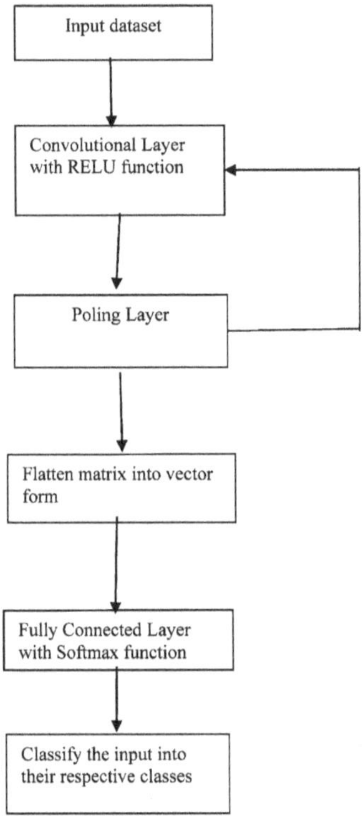

FIGURE 10.5 Basic convolutional neural network workflow.

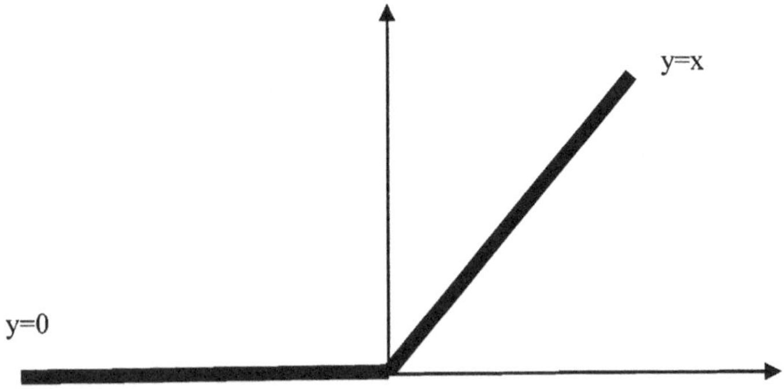

FIGURE 10.6 RELU function.

There are other activation functions are also available that are used by the various learning approaches such like sigmoid, tanh, etc. However, RELU and Softmax show the good combination with convolution network. This network achieved drastic success in image recognition or in simple terms, it works well with images. The image recognition is a powerful technique in digital forensic while analyzing the captured images of digital camera. With the purpose of authentication, unauthorized person judgment convolution model is used in this research. The next section shows the experimental work using convolution model on different three image-related datasets. Different types of image-related datasets are taken to know this neural network results on images.

10.6 EXPERIMENTAL WORK

Convolution network is one of the successful and elegant approaches of deep learning. This approach plays a vital role in the process of authentication by its image recognition ability. Because of this quality, this network is also able to analyze digital camera footage or captured images.

This study applies the convolution network on three different image datasets. To know the potential of this network in terms of accuracy, different image datasets are used. The research uses MNIST, and Dogs vs Cats datasets.

This section subdivided the work into the three other subsections. In Subsection 10.6.1, MNIST dataset is used with convolution network and then analyze the output in reference to accuracy and loss. The next Subsection 10.6.2 shows the experimental result gained by applying convolution network on Dogs vs Cats dataset. All the results gained are represented in tabular as well as in the graph form.

10.6.1 CONVOLUTIONAL NEURAL NETWORK WITH MNIST DATASET

MNIST is a digit dataset from 0 to 9 in the form of images. The size of the individual image is 28 by 28. The complete dataset is divided into training as testing phases. During training 785 columns and in testing 784 columns are present.

On applying this digit image dataset to convolution network during the training phase, the accuracy obtained with loss is as follows in Table 10.3.

TABLE 10.3 MNIST Dataset on CNN Deep Learning Model (During Training Phase).

Epochs	Accuracy	Loss
1	0.193	2.1768
2	0.3289	1.8864
3	0.4043	1.7031
4	0.4453	1.5904
5	0.4776	1.516
6	0.4971	1.4638
7	0.5125	1.417
8	0.5286	1.3768
9	0.5338	1.3594
10	0.549	1.3229

Table 10.3 shows the accuracy and loss gained during training phase using 10 epochs. Similarly, Table 10.4 shows the outcome during testing phase.

TABLE 10.4 MNIST Dataset on CNN Deep Learning Model (During Testing Phase).

Epochs	Accuracy	Loss
1	0.5083	1.6791
2	0.6386	1.1453
3	0.8252	0.7642
4	0.8683	0.603
5	0.8805	0.5196
6	0.8936	0.437
7	0.9064	0.3787
8	0.9145	0.3517
9	0.925	0.3168
10	0.9271	0.2966

Table 10.4 shows that 92.71% accuracy the convolution model gets while working on MNIST dataset. The graphical representation is depicted in Figure 10.6 in training.

Figure 10.7 shows how accuracy and loss varies with epochs during training. Similarly, Figure 10.8 shows effect of epochs on accuracy and loss during testing phase.

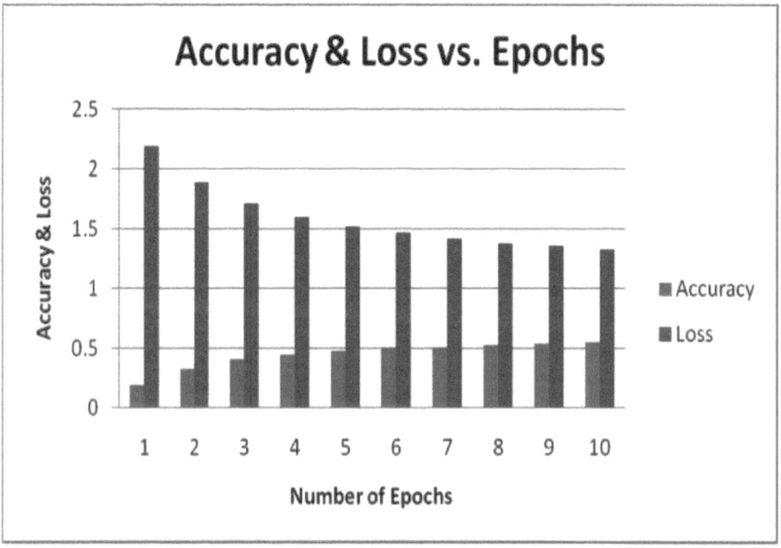

FIGURE 10.7 Accuracy and loss vs epochs (during training phase) using MNIST dataset.

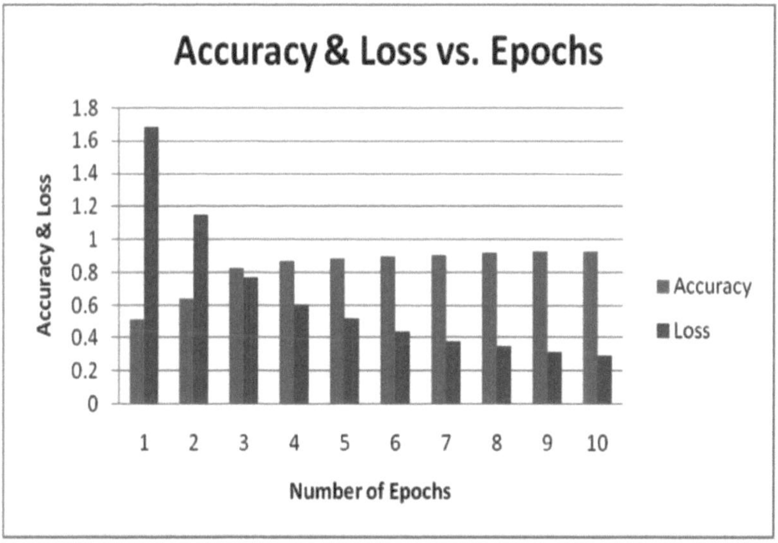

FIGURE 10.8 Accuracy and loss vs epochs (during testing phase) using MNIST dataset.

Figure 10.8 shows the impact of epochs on accuracy and loss. CNN is able to get more than 92% accuracy.

10.6.2 CONVOLUTIONAL NEURAL NETWORK WITH DOGS VS CATS DATASET

Dogs vs Cats dataset consists of 25,000 images including both dogs as well as cats. This dataset is divided into train and test datasets. During training with eight epochs, both accuracy and loss results are shown in Table 10.5.

TABLE 10.5 Dogs vs Cats Dataset on CNN Deep Learning Model (During Training Phase).

Epochs	Accuracy	Loss
1	0.6291	0.7514
2	0.7241	0.5521
3	0.7627	0.5015
4	0.7775	0.4718
5	0.8007	0.4423
6	0.8096	0.4233
7	0.8334	0.377
8	0.8439	0.3606

Figure 10.5 shows that after eight epochs more than 84% accuracy is gained during the training phase. In the same way, the Table 10.6 given below shows accuracy and loss gained during testing phase with eight epochs.

TABLE 10.6 Dogs vs Cats Dataset on CNN Deep Learning Model (During Testing Phase).

Epochs	Accuracy	Loss
1	0.6603	0.632
2	0.7749	0.4828
3	0.7535	0.4962
4	0.8371	0.397
5	0.784	0.4924
6	0.8233	0.413
7	0.8548	0.3563
8	0.8552	0.3507

Figure 10.6 shows that more than 85% accuracy is gained by CNN model on this image dataset during testing phase. Figure 10.9 graphically represents the result in training.

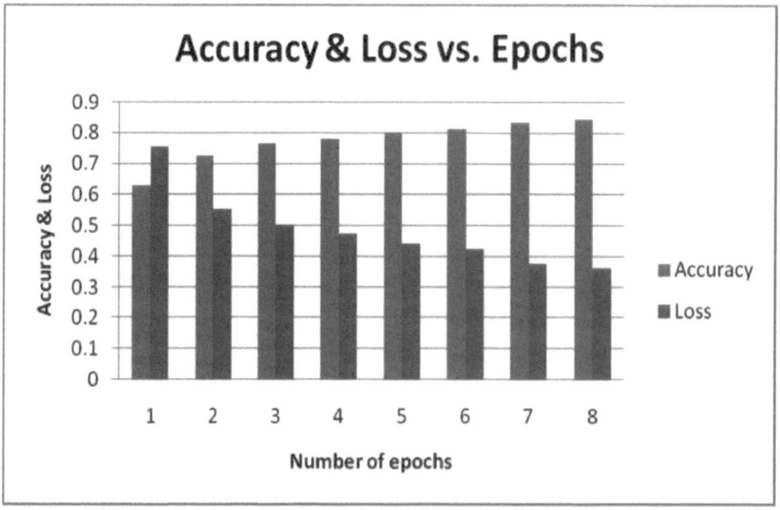

FIGURE 10.9 Accuracy and loss vs epochs (during training phase) using Dogs vs Cats Dataset.

Figure 10.9 shows graphically the gained accuracy and loss in training phase. In the same fashion, Figure 10.10 shows in testing phase.

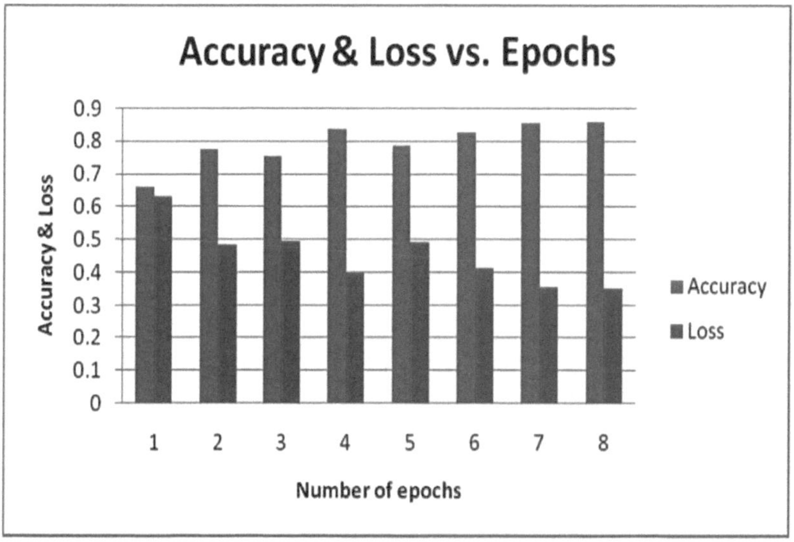

FIGURE 10.10 Accuracy and Loss vs Epochs (during testing phase) using Dogs vs Cats dataset.

Figure 10.10 shows the change in accuracy and loss with respect to epochs using Dogs vs Cats dataset with CNN.

10.7 CONCLUSION

The document concluded that convolution network is the good classifier to work on images. The chapter explains the concept of deep learning, forensic analysis, and how deep learning approach called convolutional model can be used in digital forensic analysis. The experiments performed in the study used the two well-known image datasets, namely MNIST and Dogs vs Cats datasets. During the testing phase, more than 92 and 85% accuracy is gained by the convolution model with respective datasets. To analyze the captured images in digital camera, deep learning-based convolution network model can be a good choice. In future, analyses of videos would be done using this learning model.

KEYWORDS

- **forensic**
- **security**
- **deep learning**
- **machine learning**
- **convolutional neural network**

REFERENCES

1. Mayer, O.; Stamm, M. C. Forensic Similarity for Digital Images. *IEEE Trans. Inform. Forensics Secur.* **2019,** *15*, 1331–1346.
2. van Staden, W. J. C.; van der Poel, E. Using Automated Keyword Extraction to Facilitate Team Discovery in a Digital Forensic Investigation of Electronic Communications. *SAIEE Afr. Res. J.* **2017,** *108*(2), 45–55.
3. da Cruz Nassif, L. F.; Hruschka, E. R. Document Clustering for Forensic Analysis: An Approach for Improving Computer Inspection. *IEEE Trans. Inform. Forensics Secur.* **2012,** *8*(1), 46–54.

4. Zhong, G.; Zhang, K.; Wei, H.; Zheng, Y.; Dong, J. Marginal Deep Architecture: Stacking Feature Learning Modules to Build Deep Learning Models. *IEEE Access* **2019,** *7*, 30220–30233.
5. Shao, L.; Wu, D.; Li, X. Learning Deep and Wide: A Spectral Method for Learning Deep Networks. *IEEE Trans. Neural Netw. Learn. Syst.* **2014,** *25*(12), 2303–2308.
6. Roth, A. L.; Ungvarsky, E. J. Forensic Identification and Criminal Justice: Forensic Science, Justice and Risk, by Carole McCartney. *Law Probab. Risk* **2009**, *8*, 55.
7. Ligertwood, A.; Edmond, G. Expressing Evaluative Forensic Science Opinions in a Court of Law. *Law Probab. Risk* **2012,** *11*(4), 289–302.
8. Pun, C. M.; Yuan, X. C.; Bi, X. L. Image Forgery Detection Using Adaptive Oversegmentation and Feature Point Matching. *IEEE Trans. Inform. Forensics Secur.* **2015,** *10*(8), 1705–1716.
9. Yao, H.; Wang, S.; Zhao, Y.; Zhang, X. Detecting Image Forgery Using Perspective Constraints. *IEEE Signal Process. Lett.* **2011,** *19*(3), 123–126.
10. Li, H.; Luo, W.; Qiu, X.; Huang, J. Image Forgery Localization via Integrating Tampering Possibility Maps. *IEEE Trans. Inform. Forensics Secur.* **2017,** *12*(5), 1240–1252.
11. Nguyen, C.; Redinbo, G. R. Fault Tolerance Design in JPEG 2000 Image Compression System. *IEEE Trans. Dependable Secure Comput.* **2005,** *2*(1), 57–75.
12. Keysers, D.; Deselaers, T.; Gollan, C.; Ney, H. Deformation Models for Image Recognition. *IEEE Trans. Pattern Anal. Mach. Intell.* **2007,** *29*(8), 1422–1435.
13. Galbally, J.; Marcel, S.; Fierrez, J. Image Quality Assessment for Fake Biometric Detection: Application to Iris, Fingerprint, and Face Recognition. *IEEE Trans. Image Process.* **2013,** *23*(2), 710–724.
14. Al-Dhaqm, A.; Abd Razak, S.; Othman, S. H.; Ali, A.; Ghaleb, F. A.; Rosman, A. S.; Marni, N. Database Forensic Investigation Process Models: A Review. *IEEE Access* **2020,** *8*, 48477–48490.
15. Biem, A. Minimum Classification Error Training for Online Handwriting Recognition. *IEEE Trans. Pattern Anal. Mach. Intell.* **2006,** *28*(7), 1041–1051.
16. He, R.; Wu, X.; Sun, Z.; Tan, T. Wasserstein CNN: Learning Invariant Features for NIR-VIS Face Recognition. *IEEE Trans. Pattern Anal. Mach. Intell.* **2018,** *41*(7), 1761–1773.
17. Lee, Y. K.; Chen, L. H. High Capacity Image Steganographic Model. *IEE Proc. Vision Image Signal Process.* **2000,** *147*(3), 288–294.
18. Chepushtanova, S.; Kirby, M.; Peterson, C.; Ziegelmeier, L. In *An Application of Persistent Homology on Grassmann Manifolds for the Detection of Signals in Hyperspectral Imagery*, 2015 IEEE International Geoscience and Remote Sensing Symposium (IGARSS); IEEE, 2015; pp 449–452.
19. Sousedik, C.; Busch, C. Presentation Attack Detection Methods for Fingerprint Recognition Systems: A Survey. *Iet Biom.* **2014,** *3*(4), 219–233.
20. Chugh, T.; Cao, K.; Zhou, J.; Tabassi, E.; Jain, A. K. Latent Fingerprint Value Prediction: Crowd-Based Learning. *IEEE Trans. Inform. Forensics Secur.* **2017,** *13*(1), 20–34.
21. Ali, Z.; Imran, M.; Alsulaiman, M. An Automatic Digital Audio Authentication/Forensics System. *IEEE Access* **2017,** *5*, 2994–3007.

22. Le, N.; Retraint, F. An Improved Algorithm for Digital Image Authentication and Forgery Localization Using Demosaicing Artifacts. *IEEE Access* **2019,** *7*, 125038–125053.
23. Bkassiny, M.; Li, Y.; Jayaweera, S. K. A Survey on Machine-Learning Techniques in Cognitive Radios. *IEEE Commun. Surv. Tutor.* **2012,** *15*(3), 1136–1159.
24. Krummenacher, G.; Ong, C. S.; Koller, S.; Kobayashi, S.; Buhmann, J. M. Wheel Defect Detection with Machine Learning. *IEEE Trans. Intell. Transp. Syst.* **2017,** *19*(4), 1176–1187.
25. Liu, Q.; Li, P.; Zhao, W.; Cai, W.; Yu, S.; Leung, V. C. A Survey on Security Threats and Defensive Techniques of Machine Learning: A Data Driven View. *IEEE Access* **2018,** *6*, 12103–12117.
26. Ong, Y. S.; Gupta, A. AIR 5: Five Pillars of Artificial Intelligence Research. *IEEE Trans. Emerg. Topics Comput. Intell.* **2019,** *3*(5), 411–415.
27. Basallo, Y. A.; Senti, V. E.; Sanchez, N. M. Artificial Intelligence Techniques for Information Security Risk Assessment. *IEEE Latin Am. Trans.* **2018,** *16*(3), 897–901.
28. Ardekani, B. A.; Kershaw, J.; Braun, M.; Kanuo, I. Automatic Detection of the Mid-Sagittal Plane in 3-D Brain Images. *IEEE Trans. Med. Imaging* **1997,** *16*(6), 947–952.
29. Frolov, A. A.; Husek, D.; Muraviev, I. P.; Polyakov, P. Y. Boolean Factor Analysis by Attractor Neural Network. *IEEE Transac. Neural Netw.* **2007,** *18*(3), 698–707.
30. Xin, R.; Zhang, J.; Shao, Y. Complex Network Classification with Convolutional Neural Network. *Tsinghua Sci. Technol.* **2020,** *25*(4), 447–457.
31. Li, D.; Wang, J.; Xu, J.; Fang, X. Densely Feature Fusion Based on Convolutional Neural Networks for Motor Imagery EEG Classification. *IEEE Access* **2019,** *7*, 132720–132730.
32. Pearlmutter, B. A. Gradient Calculations for Dynamic Recurrent Neural Networks: A Survey. *IEEE Trans. Neural Netw.* **1995,** *6*(5), 1212–1228.
33. Titos, M.; Bueno, A.; García, L.; Benítez, M. C.; Ibañez, J. Detection and Classification of Continuous Volcano-Seismic Signals with Recurrent Neural Networks. *IEEE Trans. Geosci. Remote Sensing* **2018,** *57*(4), 1936–1948.
34. Vishwakarma, D. K.; Upadhyay, S. A Deep Structure of Person Re-Identification Using Multi-Level Gaussian Models. *IEEE Trans. Multi-Scale Comput. Syst.* **2018,** *4*(4), 513–521.
35. Van Lanh, T.; Chong, K. S.; Emmanuel, S.; Kankanhalli, M. S. In *A Survey on Digital Camera Image Forensic Methods*, 2007 IEEE International Conference on Multimedia and Expo; IEEE, 2007; pp 16–19.
36. Koroniotis, N.; Moustafa, N.; Sitnikova, E. Forensics and Deep Learning Mechanisms for Botnets in Internet of Things: A Survey of Challenges and Solutions. *IEEE Access* **2019,** *7*, 61764–61785.
37. Štern, D.; Payer, C.; Giuliani, N.; Urschler, M. Automatic Age Estimation and Majority Age Classification from Multi-Factorial MRI Data. *IEEE J. Biomed. Health Inform.* **2018,** *23*(4), 1392–1403.
38. Rehman, A.; Naz, S.; Razzak, M. I.; Hameed, I. A. Automatic Visual Features for Writer Identification: A Deep Learning Approach. *IEEE Access* **2019,** *7*, 17149–17157.
39. Galea, C.; Farrugia, R. A. Forensic Face Photo-Sketch Recognition Using a Deep Learning-Based Architecture. *IEEE Signal Process. Lett.* **2017,** *24*(11), 1586–1590.

40. Ding, X.; Chen, Y.; Tang, Z.; Huang, Y. Camera Identification Based on Domain Knowledge-Driven Deep Multi-Task Learning. *IEEE Access* **2019,** *7*, 25878–25890.
41. Xiao, J.; Li, S.; Xu, Q. Video-Based Evidence Analysis and Extraction in Digital Forensic Investigation. *IEEE Access* **2019,** *7*, 55432–55442.
42. Khan, M. J.; Khan, H. S.; Yousaf, A.; Khurshid, K.; Abbas, A. Modern Trends in Hyperspectral Image Analysis: A Review. *IEEE Access* **2018,** 6, 14118–14129.
43. Hosler, B. C.; Zhao, X.; Mayer, O.; Chen, C.; Shackleford, J. A.; Stamm, M. C. The Video Authentication and Camera Identification Database: A New Database for Video Forensics. *IEEE Access* **2019,** 7, 76937–76948.
44. Castiglione, A.; Cattaneo, G.; De Maio, G.; De Santis, A. Automated Production of Predetermined Digital Evidence. *IEEE Access* **2013,** 1, 216–231.
45. Rekhis, S.; Boudriga, N. A System for Formal Digital Forensic Investigation Aware of Anti-Forensic Attacks. *IEEE Trans. Inform. Forensics Secur.* **2011,** *7*(2), 635–650.
46. Zhu, Y. Attack Pattern Discovery in Forensic Investigation of Network Attacks. *IEEE J. Sel. Areas Commun.* **2011,** *29*(7), 1349–1357.
47. Battisha, M.; Elmaghraby, A.; Meleis, H.; Samineni, S. Adaptive Tracking of Network Behavioral Signals for Real Time Forensic Analysis of Service Quality Degradation. *IEEE Trans. Netw. Serv. Manag.* **2008,** *5*(2), 105–117.
48. Liu, A.; Fu, H.; Hong, Y.; Liu, J.; Li, Y. $ LiveForen $: Ensuring Live Forensic Integrity in the Cloud. *IEEE Trans. Inform. Forensics Secur.* **2019,** *14*(10), 2749–2764.
49. Quick, D.; Choo, K. K. R. IoT Device Forensics and Data Reduction. *IEEE Access* **2018,** *6*, 47566–47574.
50. Guillén, J. H.; del Rey, A. M.; Casado-Vara, R. Security Countermeasures of a SCIRAS Model for Advanced Malware Propagation. *IEEE Access* **2019,** *7*, 135472–135478.
51. Marinai, S.; Gori, M.; Soda, G. Artificial Neural Networks for Document Analysis and Recognition. *IEEE Trans. Pattern Anal. Mach. Intell.* **2005,** *27*(1), 23–35.
52. Chen, M.; Challita, U.; Saad, W.; Yin, C.; Debbah, M. Artificial Neural Networks-Based Machine Learning for Wireless Networks: A Tutorial. *IEEE Commun. Surv. Tutor.* **2019,** *21*(4), 3039–3071.

Index

V

Y